ACTA BIOLOGICA ET MEDICA GERMANICA

AKADEMIE-VERLAG · BERLIN · BAND 17 · HEFT 1 · SEITE 1–115 · 1966

AUFNAHMEBEDINGUNGEN

1. Es werden nur Arbeiten angenommen, die nicht an anderer Stelle mit demselben Inhalt veröffentlicht oder zur Veröffentlichung angeboten werden. Der Autor verpflichtet sich nach Annahme, die Arbeit an keiner anderen Stelle zu veröffentlichen.

2. Die Arbeit muß wissenschaftlich wertvoll sein. Bestätigungen bekannter Tatsachen, Versuche und Beobachtungen ohne positives Ergebnis werden, wenn überhaupt, nur in kürzester Form aufgenommen. Nicht aufgenommen werden Arbeiten referierenden Charakters, Polemiken und rein spekulative Arbeiten, falls sie nicht ganz wesentliche neue Gesichtspunkte enthalten.

3. Kurzmitteilungen für experimentelle Ergebnisse werden bei der Drucklegung zeitlich bevorzugt. Für ihren Inhalt ist ausschließlich der Autor verantwortlich.

4. Die Arbeiten müssen kurz und klar geschrieben und gegliedert sein. Problematik (Einleitung), Methodik, Befunde und Diskussion, evtl. Schlußfolgerungen sollen deutlich in Erscheinung treten. Der Arbeit soll ein Kurzreferat der wesentlichsten Ergebnisse vorausgestellt werden. Neben einer anderssprachigen erscheint auf jeden Fall eine deutsche Zusammenfassung. Die Arbeiten werden in folgenden Sprachen angenommen: Deutsch, Russisch, Englisch und Französisch.

5. Die Arbeiten werden im Sofortumbruch gesetzt; größere Korrekturen in Form von Streichungen bzw. Zusätzen sind daher in der Umbruchkorrektur nicht mehr möglich.

6. Genaue Hinweise zur Manuskriptgestaltung sind von der Redaktion der Zeitschrift anzufordern und sind unbedingt einzuhalten.

7. Manuskripte sind an die Herausgeber zu senden oder direkt an die Redaktion der Acta biologica et medica germanica. 1115 Berlin-Buch, Lindenberger Weg 70.

8. Von jeder Originalarbeit werden kostenlos 80 Sonderdrucke geliefert. Darüber hinaus können bis zu 100 Sonderdrucke gegen Berechnung bezogen werden.

Die Herausgeber

ACTA BIOLOGICA ET MEDICA GERMANICA

Herausgeber:

R. Baumann · H. Dutz · A. Graffi · H. Gummel · F. Jung · L.-H. Kettler
S. M. Rapoport

Band 17 1966 Heft 1

Acta biol. med german , Band 17, Seite 1 – 7 (1966)

Cairo University – Faculty of Science – Botany Department Giza-UAR

Studies on the respiration of Fusarium moniliforme during intoxication with inhibitors of carbohydrate metabolism

A W HARHASH

(Received 4 8. 1965)

Summary

1 The effect of different inhibitors of the enzymes concerned in carbohydrate metabolism on the respiration and metabclism of Fusarium moniliforme has been studied Inhibition of respiration as measured by oxygen uptake was found to be parallel to that as measured by carbondioxide output

2 The addition of citrate, succinate, malate or pyruvate to the mycelium stimulated the respiration

3 Each of the inhibitors has caused accumulation of both pyruvate and α-ketoglutarate in the culture media, certain differences in the effect on these keto acids accumulation were noted among these inhibitors

4 Fluoroacetate had caused the accumulation of citric acid in the mycelium of the fungus

5 These results are taken to mean that at least 40% of the oxygen uptake involved the tricarboxylic acid cycle

Introduction

Several reports have been made on the reagents which inactivate sulfhydryl enzymes. Thus the effects of fluorides have been studied by LINDGREN and HARVEY [9] and VERRAL [15], of fluoroacetate by ROTHERY, BROWN and BOULTER [13] and JAMES [7], [8], of arsenite by MÜLLER-STOLL [10], of iodoacetate by NEILANDS and STUMPF [11] and of malonate by GODZESKI and STONE [5]

In animal tissues the functioning of the tricarboxylic acid cycle is well established and there is good evidence that it is at least functional in yeast and higher plants. The enzymes of the tricarboxylic acid cycle are of quite general occurrence among fungi such as Neurospora crassa (ROTHERY et al [13]), Penicillium chrysogenum (OLSON [12]) and others.

For practical reasons most of our information on the response of fungi to toxic agents relates to the evidence of inhibition of spore germination or mycelial growth. However, inhibition of respiratory activity as measured by oxygen uptake is often parallel to growth inhibition. We may also speculate that a compound which acts on respiration and growth and causes the accumulation of certain compounds in the fungal tissue is a general enzyme poison.

The object of this work is to test the importance of the citric cycle in tissue respiration and metabolism of Fusarium moniliforme.

Experimental

Seven days old fungal mats, previously grown on Dox's liquid medium, were supplied with fresh Dox's medium but with ammonium nitrate as a nitrogen source. Some of the cultures were used as controls while the others were supplied with each of sodium malonate, fluoroacetate, sodium arsenite or sodium iodoacetate to give a final concentration of 0.01 M, 5 mM, 0.01 M and 0.1 mM alternatively. The conical flasks were then stoppered with rubber stoppers through which passed inlet and outlet glass tubes which served for the exchange of gases during carbon dioxide estimation. At the beginning of each experiment and also after two intervals of 24 hours, duplicate samples of mycelial mats were filtered off and their fresh and dry weights were determined, while the filtrate was taken for analysis.

Sucrose analysis

For the estimation of the direct reducing value and total reducing value of sugar, the copper method of Schaffer and Hartmann [14] was used with the modification by Maskell and co-workers (unpublished work).

Keto acid analysis

For determination of keto acids in the culture media or the fungal mats, the method of Friedemann and Haugen [4] was followed. Chromatographic separation and estimation of pyruvic and α-ketoglutaric acids was carried out as described by El-Hawary and Thompson [2].

Estimation of citric acid

The fungal mycelium was ground with acid-washed sand and hot water (60°), and the amount of citric acid in the extract was determined by the method of Ettinger, Goldbaum and Smith [3].

Carbon dioxide estimation

The carbon dioxide outputs by the mycelial mats were determined by a procedure described by Harhash [6].

Oxygen uptake

Spore inocula of the fungus were grown on Dox's agar in a minimal medium with and without the four inhibitors. The carbon source was 0.1% sucrose. During the growth period the medium was vigorously agitated. This method of growth gave lower rates of endogenous respiration than material grown with the more usual sucrose concentration of 2%.

In some experiments 0.02 M malate, succinate or citrate was added to mycelium which had been grown in the above manner for 48 hours and the oxygen uptake was then determined.

Results

The control medium was initially acidic (i.e. *p*H 4.0) and the addition of the small amounts of the different inhibitors to such a medium did not affect its *p*H value. However, during the experimental period which lasted 48 hours there was a slight rise in the *p*H value of the culture media of the control and those containing malonate while the *p*H of the media containing fluoroacetate, arsenite or iodoacetate was not changed at all during the whole experimental period.

Table 1

The effect of various inhibitors upon the carbon dioxide output and oxygen uptake by the mats of Fusarium moniliforme. Temperature 30 °C; CO_2 output in mgms per mat; O_2 uptake in μl per 10 mg dry mycelium per hour

Treatment	Final *p*H	CO_2 estimation		O_2 estimation	
		output in 48 hours	% inhibition	uptake per hour	% inhibition
Control	5.1	121	—	380	—
0.01 M Malonate	5.0	95	21	289	24
5 mM Fluoroacetate	4.0	72	40	228	40
0.01 M Arsenite	4.0	72	40	228	40
0.1 mM Iodoacetate	4.0	72	40	220	42

Table 1 shows that the addition of the different inhibitors induced lower rates of carbon dioxide output, during the 48 hours of the experiment, as compared with the controls. Malonate inhibited the respiration of the mycelium to 21% while fluoroacetate, arsenite and iodoacetate inhibited the respiration by 40%. The results of measuring the rate of respiration in terms of μl oxygen uptake per 10 mg dry mycelium per hour seems to substantiate the results of measuring the carbon dioxide output. So it is interesting to show that fluoroacetate, arsenite and iodoacetate inhibited the oxygen uptake of the mycelial mats about 40% and that malonate inhibited the oxygen uptake 24%.

Analysis of the various culture media for sugar content: Since all the culture media initially contained sucrose with no direct reducing value then it seems that as the sucrose of the media came in contact with the fungal mats it was apparently hydrolysed into its reducing hexose components before sugar absorption; the rate of hydrolysis being faster than sugar absorption and hence the accumulation of the reducing sugars in the media (Table 2). It is interesting to note that the amount of hexoses in each of the culture media is equal to the amount of total sugars in the corresponding medium denoting

Table 2

Level of sugar in the media of normal and poisoned fungal mats of Fusarium moniliforme after 48 hours from the beginning of the experiment (Mean values are expressed in mg/50 ml)

Treatment	Sugar analysis			
	DRV[1]	TRV[2]	uptake	% decrease in uptake
Initial for all treatments	0	425	—	—
Control	132	137	288	—
0.01 M Malonate	190	190	235	20
5 mM Fluoroacetate	252	252	173	40
0.01 M Arsenite	281	281	144	50
0.1 mM Iodoacetate	281	281	144	50

[1] Direct reducing value. — [2] Total reducing value

Table 3

Level of keto acids in the media of normal and poisoned fungal mats of Fusarium moniliforme after 48 hours from the beginning of the experiment (Mean values are expressed in mg/50 ml)

Treatment	Keto acid analysis		
	Pyruvic acid	α-Ketoglutaric acid	Total keto acids
Control	0.0	0.0	0.0
0.01 M Malonate	2.2	2.4	4.7
5 mM Fluoroacetate	6.4	1.0	7.2
0.01 M Arsenite	3.6	3.5	7.2
0.1 mM Iodoacetate	5.5	1.5	7.0

that the different inhibitors, used in the experiment, had no effect on the activity of the specific sucrose hydrolysing enzyme.

Comparing the sugar uptake of the culture media treated with the different inhibitors with that of the control it could be seen that malonate decreased sugar uptake 20%, fluoroacetate 40% and arsenite or iodoacetate inhibited the uptake of sugar by the mycelial mats 50%.

The results in table 3 clearly show that during the 48 hours of the experiment the experimental culture media supplied with the different inhibitors contained some keto acids apparently being produced by and diffused from the fungal mats. The individual keto acids hydrazones were separated chromatographically and the analysis showed that they were combination of pyruvic and α-ketoglutaric acids. The pyruvic acid level in the fluoroacetate and iodoacetate containing media was found to be much greater than the α-ketoglutaric level, while in the malonate and arsenite containing media the level of both keto acids was the same.

Since the appearance of pyruvic acid in the medium may be expected to depend on the rate of formation of pyruvate by the fungal mats as well as on its rate of utilization, it may be argued that the experiment demonstrated that these inhibitors have partially or wholly blocked pyruvate metabolism in the fungal tissue. The effect of fluoroacetate, arsenite or iodoacetate on blocking pyruvate metabolism was comparatively more than that of malonate.

When the fungal mats were tested for their keto acid contents they showed the presence of only traces of such keto acids. It seems therefore, that the keto acids accumulated during the metabolic activities of the fungal mats were excreted from the mats into their corresponding media.

Table 4

The accumulation of citric acid by the mycelium of Fusarium moniliforme in the presence of fluoroacetate and malonate after 48 hours from the beginning of the experiment

Treatment	Dry weight in grams of mycelium	mg citric acid/g dry weight mycelium
Control	0.42	2.14
Fluoroacetic acid	0.26	18.42
Malonic acid	0.32	9.84

For studying the accumulation of citric acid by the mycelium of Fusarium moniliforme in the presence of fluoroacetate or malonate, cultures of the fungus were grown in liquid medium for 20 hours. Fluoroacetic acid (10^{-2} M) or malonic acid (10^{-1} M) was added to some cultures and the whole cultures were grown for further 30 hours. The results given in table 4 show that citric acid accumulated in the mycelium when grown in the presence of fluoroacetate and malonate. The amount of citric acid accumulated in case of fluoroacetate was nearly two times more than that in case of malonic acid.

Discussion

Four inhibitors known to inhibit various enzymes of the tricarboxylic acid cycle were used at a concentration and *p*H value which assured maximum effectiveness and selectivity of action (James [8]). It could be noticed that the respiration of the whole cells as measured in terms of gas exchange due to the presence of the different inhibitors varied from 20—40%. The addition of citrate, succinate, malate or pyruvate to the mycelium stimulated the respiration.

In this connection it is worthy to mention that Yeoman [16] has reported that malonate penetrated into the cells of Aspergillus niger but did not exert its normal effect as a competitive inhibitor for the TCA cycle. In the present work malonate decreased the rate of respiration nearly 50% less than the other inhibitors. It may be then suggested that some of the malonate which penetrated the cells was stored within the mycelium. Alternatively, the penetration of malonate was relatively retarded through the cell membranes of the fungus.

The amount of pyruvate accumulated in the media in the presence of fluoroacetate or iodoacetate was much higher than that of α-ketoglutarate showing that the blocking of pyruvate was in a stage prior to α-ketoglutarate formation. This result was found to be parallel to the more accumulation of citric acid in the mycelium treated with fluoroacetate than in that treated with malonic acid. The presence of a good deal of α-ketoglutarate in the media containing malonate resulted probably from the inhibition of the succinic dehydrogenase, causing a diminution in the rate of α-ketoglutarate breakdown. Arsenite caused also the inhibition of the oxidative decarboxylation system of α-ketoglutarate leading to the accumulation of this intermediate.

The accumulation of citric acid in the presence of fluoroacetate showed the inhibition of respiration to be due to the inhibition of aconitase in this instance.

In conclusion it may be stated that these results are taken to mean that at least 40% of the oxygen uptake involved the tricarboxylic acid cycle. The fact that the TCA cycle has been shown to be involved in oxygen uptake to a large extent in Allomyces (Bonner and Machlis [1]), in Neurospora crassa (Rothery et al. [13]) and also in Fusarium moniliforme (genera takonomically unrelated) suggests that this phenomenon may be of importance in fungi generally.

Acknowledgement

The author records his sincerest thanks to Prof. Dr. S. Rapoport, Director of Physiological Chemistry Institute — Humboldt University — Berlin, for his helpful suggestions during writing this work.

References

[1] Bonner, B. A. and L. Machlis: Plant Physiol. **32**, 291 (1957).
[2] El-Hawary, M. F. S. and R. H. S. Thompson: Biochem. J. **53**, 340 (1953).
[3] Ettinger, R. H., L. R. Goldbaum and L. H. Smith: J. biol. Chemistry **199**, 531 (1952).
[4] Friedemann, T. E. and G. E. Haugen: J. biol. Chemistry **147**, 415 (1943).
[5] Godzeski, C. and R. W. Stone: Arch. Biochem. Biophysics **59**, 132 (1955).
[6] Harhash, A. W.: Studies on the metabolism of Fusarium oxysporum, Ph. D. thesis, Cairo University (1958).
[7] James, W. O.: Plant Respiration, Oxford: Clarendon Press 1953.
[8] James, W. O.: Ann. Rev. Plant Physiol. **4**, 59 (1953).
[9] Lindgren, R. M. and M. H. Harvey: J. Forest Products Research Society **2**, 1 (1952).
[10] Müller-Stoll, W. R.: Phytopathol. Z. **17**, 265 (1950).
[11] Neilands, J. B. and P. K. Stumpf: Outlines of Enzyme Chemistry, New York, John Wiley and Sons, p. 315 (1955).
[12] Olson, J. A.: Nature [London] **174**, 695 (1954).

[13] Rothery, W. G., Brown A. W. and D. Boulter: New Phytologist **61**, 41 (1962).
[14] Schaffer, L. A. and A. F. Hartmann: J. biol. Chemistry **45**, 349 (1921).
[15] Verral, A. F.: J. agric. Res. **78**, 695 (1949).
[16] Yeoman, M. M.: New Phytologist **64**, 65 (1965).

Zusammenfassung

A. W. Harhash: Untersuchungen zur Atmung von Fusarium moniliforme bei Vergiftung mit Hemmstoffen des Kohlenhydratstoffwechsels

1. Der Einfluß verschiedener Hemmstoffe einiger am Kohlenhydratstoffwechsel beteiligter Enzyme auf die Atmung und den Stoffwechsel von Fusarium moniliforme wurde untersucht. Die nach der Sauerstoffaufnahme und nach der Kohlendioxydabgabe gemessene Atmungshemmung waren gleich.

2. Zugabe von Zitrat, Sukzinat, Malat oder Pyruvat zum Myzel stimulierte die Atmung.

3. Alle Hemmstoffe bewirkten eine Akkumulation von Pyruvat und α-Ketoglutarat in den Kulturmedien, wobei bestimmte Unterschiede in der Beeinflussung der Ketosäureakkumulation für die verschiedenen Inhibitoren festgestellt wurden.

4. Fluorazetat verursachte eine Akkumulation von Zitronensäure im Myzel.

5. Auf Grund der erhaltenen Resultate wird festgestellt, daß mindestens 40% der Sauerstoffaufnahme über den Zitronensäurezyklus erfolgt.

Резюме

А. В. Гаргаш: Исследования о дыхании мицелия Fusarium moniliforme при отравлении ингибиторами углеводного метаболизма

1. Изучали влияние разных ингибиторов некоторых энзимов, участвующих в углеводном метаболизме, на дыхание и обмен веществ Fusarium moniliforme. Торможение дыхания, замеренное по поглощению кислорода и по отдаче двуокиси кислорода, оказалось одинаковым.

2. Добавление к мицелию цитрата, сукцината, малата или пирувата стимулировало дыхание.

3. Все ингибиторы вызвали аккумулирование пирувата и α-кетоглютарата в культуральных средах, причем отмечались определенные различия во влиянии отдельных ингибиторов на аккумулирование кетокислоты.

4. Ацетат фтора вызывает аккумулирование лемонной кислоты в мицелии.

5. Полученные результаты показывают, что не менее 40% поглощения кислорода осуществляется через цикл лемонной кислоты.

Acta biol. med. german., Band 17, Seite 8–16 (1966)

Cairo University, Faculty of Science, Botany Department, Giza-UAR

Studies on the effect of various carbohydrate and inorganic nitrogen sources on the metabolism of Fusarium moniliforme

A. W. HARHASH

(Received 23. 8. 1965)

Summary

The assimilation of some inorganic nitrogen sources by Fusarium moniliforme has been studied. The results showed that the fungus absorbed ammonium ions in preference to nitrate ions whether supplied separately or together in the form of ammonium nitrate.

The mycelium assimilated glucose more rapidly than fructose or L-arabinose. The results are taken to indicate that utilization of fructose or L-arabinose is dependent on an enzyme system which is adaptive in this fungus.

Keto acid production showed the presence of pyruvic and α-ketoglutaric acids in the culture media containing sodium nitrate and glucose. When glucose was substituted by fructose or L-arabinose the culture media contained an insignificant amount of keto acids.

Introduction

Although in many cases our knowledge of the action of the nitrogen compounds is well founded, no single pattern of nitrogen can be described for all fungi. In general, however, nitrates have often been considered as excellent sources of nitrogen for many fungi (HACSKAYLO, LILLY and BARNETT [3] and SACKSENA et al. [11]). An inorganic pathway of nitrate reduction to nitrite then to ammonia has been proposed by SILVER and MCELROY [14].

According to ROBBINS [9] no instances have been recorded in the literature of an organism being able to utilize nitrate nitrogen and unable to utilize ammonium nitrogen. On the other hand, the preferential utilization of ammonium ions from a medium containing ammonium nitrate has been often shown (MACMILLAN [5] and MORTON [7]).

Glucose is biologically the most important sugar and is utilized for growth by virtually all cultivable fungi, although few forms have been reported not to grow with glucose (SCHADE [12]). According to FRUTON and SIMMONDS [2] it seems possible that the ability of an organism to use fructose or L-arabinose depends upon its ability to convert the sugar in question into a phosphorylated derivative of glucose.

In the work, here described the absorption of nitrogen by Fusarium moniliforme from various nitrate and ammonium sources, and that of sugar from various hexose and pentose sources is compared. The keto acid production by this fungus in the presence of these different nitrogen and carbohydrate sources was also estimated.

Material and Methods

The fungus used in the present work was a Fusarium moniliforme. In all the experiments the fungal spores were grown in conical flasks of 150 ccs capacity, each containing 25 ml. Dox's medium in which cane sugar is used as carbon source and potassium nitrate as nitrogen source. Culture flasks were then incubated at 25 °C for seven days after which the media were decanted and the mycelial mats were washed with sterile distilled water, and the washings decanted as far as possible. The mats were subsequently kept in the culture flasks ready to be supplied with fresh Dox's medium with substituted carbon and nitrogen sources. Duplicate samples were always made for each treatment. At the beginning of each experiment, and also after two intervals of 24 hours duplicate samples of mycelial mats were filtered off and their fresh and dry weights were determined, while the filtrate was taken for analysis.

Nitrogen analysis

For ammonia determination, the sample was mixed with 10% suspension of magnesium oxide in water and distilled under reduced pressure. The ammonia produced was collected in N/70 HCl and the excess of the acid was titrated with N/70 NaOH solution.
The nitrate nitrogen was estimated by reduction to ammonia with reduced iron, and then distilled in a modified "Micro-KJELDAHL" (modification of PARNAS and WAGNER) apparatus in the usual way.

Carbohydrate analysis

For the estimation of the direct reducing sugars, the SCHAFFER-HARTMANNS copper method [13] was used after being modified by MASKELL and co-workers (unpublished work).
Determination of fructose and arabinose was carried out by both the copper method and colorimetric method of ROE [10] for fructose and MEIJBAUM [6] for arabinose, and the results were always concordant.

Keto acids analysis

For determination of keto acids in the culture media or the fungal mats, the technique used by the author in a previous work (HARHASH [4]) was adopted.

Results

pH values

The basal medium used was initially acidic (i.e. *p*H 5.5). The *p*H value of the basal medium alone and that with glucose but without nitrogen source was not changed at all during the whole experimental period, while the media containing potassium nitrate showed a rise in their *p*H values unaffected by the absence or the presence of the different carbon sources (Table 1). On the other hand, the basal media which contained ammonium nitrate or ammonium sulphate showed a steep drop in their *p*H values, the drop being steeper in the presence of the latter. The rise in the *p*H values might be ascribed by the absorption of anions while the drop might be due to the absorption of cations of the different nitrogen sources.

Table 1

Changes in the *p*H values of the various culture media

Treatment	*p*H values after	
	24 hours	48 hours
Basal medium[1] + gluc. + pot. nitrate	5.7	6.0
Basal medium + fruct. + pot. nitrate	5.7	6.0
Basal medium + arabin. + and nitrate	5.7	6.0
Basal medium + gluc. + ammonium sulphate	4.5	3.8
Basal medium + gluc. + ammonium nitrate	5.0	4.5
Basal medium without N and C sources	5.5	5.5
Basal medium + pot. nitrate	5.8	6.0
Basal medium + gluc.	5.5	5.5
Water	6.7	6.7

[1] The basal medium was Dox's liquid medium without carbon and nitrogen sources

Sugar utilization

It may be seen from table 2 that the fungal mats were able to utilize glucose, fructose and L-arabinose. The amount utilized was dependent on the kind of the sugar and the nitrogen sources used with it.

Table 2

Uptake of sugar from the various culture media
(calculated as mg glucose per 25 ml medium)

Carbon and nitrogen sources added to the basal medium[1]	Glucose or fructose or L-arabinose utilized in		
	1 st 24 hours	2 nd 24 hours	Total 48 hours
Glucose + pot. nitrate	120	177	297
Fructose + pot. nitrate	30	136	166
L-Arabinose + pot. nitrate	33	135	168
Glucose + ammonium nitrate	90	208	298
Glucose + ammonium sulphate	65	150	215
Glucose without N source	105	150	255

[1] The basal medium was Dox's liquid medium without carbon and nitrogen sources

Comparing the amounts of sugars absorbed by the mats having potassium nitrate as a nitrogen source, it appears that the greatest sugar absorption during the first 24 hours was from glucose, being four times greater than from fructose or L-arabinose. During the second 24 hours there was still more absorption from glucose, but the amount absorbed was only 1.3 times that of each of the two other sugars.

It appears from this table also that in the first 24 hours the fungal mats grown on media containing potassium nitrate absorbed more glucose than the fungal mats grown on media containing either ammonium nitrate or ammonium sulphate. However, the fungal mats absorbed practically all the remaining glucose in the media containing potassium nitrate or ammonium nitrate during the second 24 hours, while the absorption was still retarded in the media containing ammonium sulphate. This might be either due to the continued steep drop in the *p*H values of these latter media, or the nitrate anions apparently induced more absorption from glucose than did the sulphate anions.

Nitrogen uptake

It appears clearly from table 3 that when the ammonium cation was supplied together with nitrate anion, the nitrate utilization was completely stopped and the fungus selectively absorbed all its nitrogen from the ammonium ions.

Table 3

Nitrate- and ammonium-nitrogen utilization by the fungal mats calculated as mg N per 25 ml medium (i.e. per mat)

Nitrogen and carbon sources	Nitrate- or ammonia- N utilized in		
	1 st 24 hours	2 nd 24 hours	Total 48 hours
Glucose + pot. nitrate	1.0	1.6	2.6
Glucose + ammonium sulphate	4.0	2.0	6.0
Glucose + ammonium nitrate: Nitrate-N	—	—	—
Glucose + ammonium nitrate: Ammonium-N	2.1	2.0	4.1
Fructose + pot. nitrate	1.2	1.2	2.4
Arabinose + pot. nitrate	1.2	1.0	2.2
Potassium nitrate and no carbon sources	0.60	0.4	1.0

Comparing the amounts of nitrogen absorbed by the fungus from ammonium nitrate and ammonium sulphate, it appears that more ammonium ions were absorbed from the latter than from the former during the first 24 hours. However, considering that the amount of ammonium nitrogen initially supplied in the form of ammonium nitrate was half that supplied in the form of ammonium sulphate then one might be tempted to conclude that this higher absorption from ammonium sulphate was due to the initial concentration of the ammonium ions in the medium. The retardation in the absorption of the ammonium ions from ammonium sulphate in the second 24 hours might be due to the steep decrease in the *p*H of the medium during this period.

Feeding with fructose or L-arabinose as a carbon source had no effect on the absorption of nitrate ions as the amount of nitrate nitrogen absorbed by the fungal mats in the media containing either of the two sugars was nearly the same as that absorbed when glucose was the carbon source. When no carbon source was provided the fungal mats absorbed only half the amount of nitrate nitrogen absorbed from a medium containing a carbon source.

Keto acid analysis

It appears from table 4 that the largest amount of keto acids were recovered from the media containing sodium nitrate as source of nitrogen and glucose as source of carbon. When glucose was substituted by fructose or L-arabinose very little or practically no keto acids were produced.

Table 4
Keto acids analysis in the culture media
(as mg pyruvic acid per 25 ml media)

Treatment	Keto acids recovered after	
	24 hours	48 hours
Basal med. + gluc. + pot. nitrate	5.65	5.75
Basal med. + fruct. + pot. nitrate	0.17	0.32
Basal med. + L-arabin. + pot. nitrate	0.04	0.16
Basal med. + gluc. + ammonium sulphate	0.70	0.76
Basal med. + gluc. + ammonium nitrate	1.88	0.65
Basal med. + pot. nitrate	0.43	0.15
Basal med. + glucose	0.55	0.15
Basal med.	0.04	0,02
Water	0.18	0.02

When potassium nitrate was substituted by ammonium sulphate or ammonium nitrate the media showed very little keto acid content as compared with the media containing potassium nitrate. This is interesting because it indicates that nitrate absorption and utilization by the fungus specifically induced much higher production of keto acids than ammonium utilization, provided that glucose is present in the culture media.

Chromatographic analysis of the keto acids found in the various culture media of this experiment showed that they were combination of both pyruvic and α-ketoglutaric acids, the R_f value of the coloured area on the paper chromatogram occupied by the given hydrazones extracted from the media were measured and were found to be exactly equal to the R_f values of the coloured areas occupied by the hydrazones of synthesized keto acids namely pyruvic and α-ketoglutaric acids. In this connection it is of value to mention that the synthetic and natural pyruvic acid gave two spots on the chromatogram. These two spots were proved by STEWARD [15] to be the cis- and trans-forms of pyruvic acid 2:4 dinitrophenylhydrazone.

Dry weight of the fungal mats

It could be seen from table 5 that when the fungal mats were grown on basal medium without any carbon source or on distilled water, the dry weight of each mat decreased during the experimental period; this decrease being presumably due to utilization of some carbon constituents of the mat itself in respiration and other metabolic activities.

Table 5

Dry weights of the fungal mats (in mg)
(Initial dry weight of each mat was 240 mg)

Treatment	Dry weight after	
	24 hours	48 hours
Basal med. + gluc. + pot. nitrate	255	265
Basal med. + fruct. + pot. nitrate	264	265
Basal med. + arabin. + pot. nitrate	255	255
Basal med. + gluc. + ammonium sulphate	250	255
Basal med. + gluc. + ammonium nitrate	257	270
Basal med. + pot. nitrate	215	185
Basal med. + glucose	250	250
Basal medium	195	175
Water	235	200

On the other hand, addition of carbon source to the basal medium alone or together with a nitrogen source induced an increase in the dry weight during the experimental period. The slightest increase was when L-arabinose was used as a carbon source while it was relatively high when glucose together with sodium- or ammonium-nitrate were supplied to the basal medium. The small increase in the dry weight when ammonium sulphate was supplied as a source of nitrogen may have been due to the strong acidity of the culture media (*p*H 3.8).
When glucose was substituted by fructose, the latter induced the same increase in the dry weight of the fungal mats as that formed when glucose was used.

Discussion

The results showed that Fusarium moniliforme absorbed ammonium ions in preference to nitrate ions whether supplied separately or together in the form of ammonium nitrate. The fungus absorbed and utilized during the 48 hours of the experiment a good deal of the ammonium nitrogen leaving all the nitrate nitrogen in the culture media. These results support the suggestions of Brian et al. [1] that a definite antagonism exists between the metabolic pathways involved in nitrate and ammonium utilization by Myrothecium verrucaria and in the presence of ammonium nitrogen the nitrate pathway is blocked.
Ammonium utilization by the fungal mats was more from ammonium sulphate than from ammonium nitrate during the first 24 hours of the experiment and so the drop in the *p*H of the media was more in the former case than in the latter. In this connection it may be worthy to mention that the author suggested on the basis of studies on Fusarium oxysporum that the drop in *p*H may be due to the accumulation of hydrogen ions originating by exchange of ammonia ions absorbed from the media.

Comparison of the sugar uptake from media containing glucose, fructose or L-arabinose seemed to point out that glucose was more rapidly absorbed from the culture media by the fungus than fructose and L-arabinose. Although the absorption was greater from glucose than from fructose, yet the increase in the dry weight of the fungal mats supplied with fructose was slightly more than in those supplied with glucose. This might indicate that glucose was more rapidly pulled in the metabolic pool of the tricarboxylic acid cycle than fructose.

It is interesting to note also that the fungus was found to utilize small amounts from fructose and L-arabinose during the first 24 hours of the experiment; the utilization of glucose being 4 times greater than each of both other sugars. On the other hand, the fungus could increase greatly its absorption capacity from fructose and L-arabinose in the second 24 hours. The final amount absorbed from both sugars and also the final growth was comparable to that of glucose. It would seem, therefore, that utilization of fructose and L-arabinose is dependent on an enzyme system which is adaptive in the fungus.

The results obtained in this investigation showed that the highest keto acids production occured on feeding with glucose plus potassium nitrate. The produced keto acids were identified, by chromatographic analysis, and were found to be pyruvic and α-ketoglutaric acids; the former forming the major component. It was shown by Nason [8] using Neurospora crassa, Morton [7] using mould fungi and Wirth and Nord [16] using Fusarium lini that nitrate reductase enzyme was formed by those fungi when grown on nitrate medium. It was presumed that this enzyme catalyzed the reduction of nitrate to nitrite with the concurrent oxidation of the carbon source. According to these findings, it is thus probable to suggest that Fusarium moniliforme metabolized glucose by means of a system similar to the Embeden-Meyerhof pathway leading to the formation of pyruvic acid, the decarboxylation of which being partially inhibited by the presence of nitrite formed from the reduction of nitrate.

The presence of α-ketoglutaric acid in the culture media substantiate the suggestion of the author in a previous work Harhash [4] of the occurence of the well known tricarboxylic acid cycle in Fusarium moniliforme and in the present work the nitrite partially inhibited the decarboxylation of this keto acid.

Although glucose and nitrate absorption continued till the end of the experiment, yet the amount of keto acids recovered in the media remained constant during the second 24 hours of the experiment. It seems then that the amount of nitrite formed in the second 24 hours was too small to have any inhibitory effect on the carboxylase enzyme system and hence the non-accumulation of keto acids.

Keto acid analysis showed an insignificant amount of keto acids in the culture media containing potassium nitrate plus either fructose or L-arabinose. The absorption of these two sugars by the fungal mats was slow as compared with that of glucose. Therefore, taking into consideration that nitrate nitrogen was comparatively equally absorbed from the media containing each of the three sugars, it might be then assumed that fructose and L-arabinose were utilized in respiration at such a slow rate which did not lead to the accumulation of the intermediate keto acids.

Acnowledgement

The author wishes to acknowledge with thanks the helpful suggestions and illuminating discussions of Prof. Dr. S. RAPOPORT, director of the Physiological Chemistry Institute, Humboldt-University, Berlin.

References

[1] BRIAN, P. W., P. J. CURTIS and H. G. HEMMING: Proc. Roy. Soc. (London), Ser. B, **135**, 106 (1947).
[2] FRUTON, J. S. and S. SIMMONDS: General Biochemistry, New York, John Wiley and Sons, p. 940 (1953).
[3] HACSKAYLA, J., V. G. LILLY and H. L. BARNETT: Mycologia [New York] **46**, 691 (1954).
[4] HARHASH, A. W.: Acta biol. med. german **17**, 1 (1966).
[5] MACMILLAN, A.: Physiol. Plantarum [Copenhagen] **9**, 470 (1956).
[6] MEIJBAUM, W.: Hoppe-Seyler's Z. physiol. Chem., **258**, 117 (1939).
[7] MORTON, A. G.: J. exp. Bot. [London] **7**, 97 (1956).
[8] NASON, A.: In: W. D. MCELROY and B. GLAES (eds), Symposium on Inorganic Nitrogen Metabolism, p. 109, 136. Baltimore: The Johns Hopkins Press (1956).
[9] ROBBINS, W. J.: Amer. J. Bot. **24**, 243 (1937).
[10] ROE, J. H.: J. biol. Chemistry **107**, 15 (1934).
[11] SACKSENA, R. K., S. K. JAIN and S. M. H. JAFRI: J. Indian bot. Soc. **31**, 281 (1953).
[12] SCHADE, A. L.: Amer. J. Bot. **27**, 376 (1940).
[13] SCHAFFER, L. A. and A. F. HARTMANN: J. biol. Chemistry **45**, 349 (1921).
[14] SILVER, W. S. and W. D. MCELROY: Arch. Biochem. Biophysics **51**, 379 (1954).
[15] STEWARD, H. B.: Biochem. J. **55**, XXVI (1953).
[16] WIRTH, J. C. and F. F. NORD: Arch. Biochem. Biophysics **2**, 463 (1943).

Zusammenfassung

A. W. HARHASH: Untersuchungen über den Einfluß verschiedener Kohlenhydrat- und anorganischer Stickstoffverbindungen auf den Stoffwechsel von Fusarium moniliforme

Es wurde die Assimilation einiger anorganischer Stickstoffverbindungen durch Fusarium moniliforme untersucht. Ammoniumionen werden gegenüber Nitrationen bevorzugt aufgenommen, unabhängig davon, ob sie getrennt oder zusammen in Form von Ammoniumnitrat zugeführt wurden.

Das Myzel assimilierte Glukose schneller als Fruktose oder L-Arabinose. Offenbar ist der Umsatz von Fruktose oder L-Arabinose von einem adaptiven Enzymsystem abhängig.

In den Kulturmedien, die Natriumnitrat und Glukose enthalten, konnten Pyruvat und α-Ketoglutarsäure nachgewiesen werden. Bei Ersatz von Glukose durch Fruktose oder L-Arabinose enthielten die Kulturmedien nur eine unbedeutende Menge an Ketosäuren.

Резюме

А. В. Гаргаш: Изучение влияния разных углеводных и неорганических источников азота на метаболизм Fusarium moniliforme

Изучали ассимиляцию некоторых источников азота мицелием Fusarium moniliforme. Ионы аммония поглощаются скорее, чем ионы нитрата, независимо от того, подаются ли они отдельно или вместе в виде нитрата аммония. Поглощение мицелием глюкозы прошла быстрее чем поглощение фруктозы или Л-арабинозы. Обмен фруктозы или L-арабинозы очевидно зависим от адаптивной энзимной системы.

В культуральных средах, содержащих нитрат натрия и глюкозу, удалось обнаружить пируват и α-кетоглютаровую кислоту. При замене глюкозы фруктозой или L-арабинозой, культуральные среды содержали лишь незначительное количество кетоновых кислот.

Acta biol. med. german., Band 17, Seite 17—24 (1966)

Aus dem Institut für Krebsforschung der Deutschen Akademie der Wissenschaften zu Berlin, Bereich Robert-Rössle-Klinik (Direktor: Prof. Dr. H. GUMMEL), Abteilung für Chemotherapie (Leiter: Dr. habil. G. BACIGALUPO)

Feedback-Regulation der Aspartat-Transkarbamylase aus verschiedenen Geweben

E. HEISE und M. GÖRLICH

(Eingegangen am 28. 1. 1966)

Zusammenfassung

Der Einfluß von Adenosintriphosphat, Zytidintriphosphat, Uridintriphosphat und 6-Aza-Uridin auf die Aktivität der Aspartat-Transkarbamylase wurde in normaler Leber, Hepatomen und WALKER-Karzinomen untersucht, wobei Unterschiede in der Regulierbarkeit dieses Enzyms aus normalen und kanzerisierten Geweben festgestellt werden konnten. Die Aktivitätsbeeinflussung der Aspartat-Transkarbamylase durch die genannten Substanzen ist in Homogenaten aus kanzerisierten Geweben stets geringer als in normalen Leberhomogenaten. Die Aspartat-Transkarbamylase unterliegt zusätzlich einer „Pseudofeedback-Regulation" durch 6-Aza-Uridin, die sich auf alle untersuchten Gewebe erstreckt und biochemischer Ausdruck einer relativ unspezifischen Wirkung dieses Antimetaboliten ist.

Die Effektivität von Stoffwechselwegen, die sich in dem Ausmaß des Basalstoffwechsels, in der Wachstumspotenz oder aber in einer erhöhten oder erniedrigten funktionellen Leistung eines bestimmten Organs ausdrückt, ist von der Aktivität der die einzelnen Metabolite umsetzenden Enzyme abhängig. Dabei besitzen die sogenannten Schlüsselenzyme, die sehr häufig die limitierenden Glieder eines Stoffwechselweges darstellen, eine besondere Bedeutung für den Stoffumsatz. Die Aktivität dieser Enzyme ist in entscheidendem Maße von der Konzentration bestimmter Metabolite, den sogenannten allosterischen Effektoren abhängig und kann durch Veränderungen ihrer Konzentrationen variiert werden. In diesem Zusammenhang sind vor allem die Regulationsmechanismen der Feedback-Kontrolle u. a. bedeutungsvoll, da sie einen großen Einfluß auf die normale Regulation von Stoffwechselprozessen und letztlich auch des Wachstums besitzen. Damit wird aber die Beeinflußbarkeit einer Enzymaktivität durch regulierende Metabolite zu einem entscheidenden Faktor für die normale Wachstumsregulation und -kontrolle. Aus diesem Grunde erscheint es sinnvoll, die Regulierbarkeit der Aktivität von Schlüsselenzymen in normal proliferierenden Geweben mit den Regulationsmechanismen von schnell und regellos wachsenden Geweben, wie sie z. B. die Transplantationstumoren darstellen, zu vergleichen und mögliche Unterschiede in der Ansprechbarkeit eines bestimmten Schlüsselenzyms auf regulierende Metabolite mit als eine Ursache für das regellose Wachstum von Tumoren anzusehen. Zu diesen regulierbaren Schlüsselenzymen gehört auch die Aspartat-Transkarbamylase (ATC)[1], die den ersten Schritt der de novo-Synthese von Pyrimidinen, die Bildung von Ureidobernsteinsäure, katalysiert und somit auch für die Bildung von Nukleinsäuren verantwortlich ist.

Durch die Untersuchungen von GERHART und PARDEE [1] ist bekannt, daß dieses Enzym in E. coli einer Feedback-Kontrolle unterliegt, wobei CTP ein starker Inhibitor ist, während die Aktivität der ATC durch ATP schwach gesteigert wird.
Auch in der Leber konnte dieses Enzym nachgewiesen werden (CALVA und COHEN [2]). Nach diesen Autoren besteht ein direkter Zusammenhang zwischen der Aktivität der ATC und dem Wachstum in diesem Organ, da 48 Std. nach einer partiellen Hepatektomie eine Aktivitätssteigerung um 97% beobachtet werden konnte, die sich im Verlauf von 15 Tagen wieder normalisierte. Eine gleichartige, jedoch bleibende Aktivitätserhöhung konnte auch in Hepatomen festgestellt werden. Schon von COHEN wird daraufhingewiesen, daß möglicherweise die Wachstumsregulation in regenerierender Leber durch eine Feedback-Kontrolle der ATC mit verursacht wird, die in Hepatomen ganz oder teilweise durchbrochen ist. Von KIM und COHEN [3] stammen Untersuchungen über die Aktivität dieses Enzyms während der Embryonalentwicklung der Rattenleber. Diese Entwicklungsperiode ist durch einen ständigen Abfall der ATC-Aktivität gekennzeichnet, der postnatal noch einige Tage anhält und schließlich einem unteren Grenzwert zustrebt, der dann während des weiteren Lebens nicht mehr unterschritten wird.

Der offensichtlich zwischen ATC-Aktivität und Wachstum bestehende Zusammenhang führte uns zu Untersuchungen über den Einfluß verschiedener Metabolite und von 6-Aza-Uridin auf die Aktivität dieses Enzyms in normaler Leber, in Hepatomen und in transplantierten WALKER-Tumoren der Ratte in vitro.

Material und Methodik

Als Versuchstiere dienten weibliche Sprague-Dawley-Ratten, die Geschlechtsreife besaßen und mit einer Standard-Keks-Kost nach KÜSSNER (zit. bei [4]) ernährt wurden. Die Tiere erhielten Wasser ad libitum. Die Hepatome erzeugten wir durch tägliche Verfütterung von 5 mg Dimethylaminoazobenzol (Buttergelb) in Sonnenblumenöl. Zur Untersuchung gelangten nur Tiere, deren Lebern eine makroskopisch deutliche Hepatombildung zeigten. Die WALKER-Tumore wurden transplantiert und 10 bis 14 Tage nach der Transplantation verwendet.
Die *Bestimmung der ATC-Aktivität* erfolgte im Prinzip nach der Methode von KORITZ und COHEN [5], wobei jedoch eine 0,5%ige wässrige Lösung von Natrium-diphenylamin-p-sulfonat anstelle der von KORITZ angegebenen (1%ig) verwendet wurde. L-Aspartat (Reanal), Diazetylmonoxim (Koch and Light) und Kaliumpersulfat (VEB Laborchemie Apolda) wurden käuflich erworben, während Dilithium-Karbamylphosphat nach der Methode von SPEKTOR et al. [6] synthetisiert wurde. Natriumdiphenylamin-p-sulfonat wurde ebenfalls in unserem Laboratorium durch Sulfonierung von Azetyldiphenylamin hergestellt. Nach Verseifung der Azetylgruppe wurde die Sulfonsäure als Bariumsalz isoliert, mehrmals aus kochendem Wasser umkristallisiert und mit Natriumsulfat in das Natriumsalz überführt. Die Gewebepärparation

[1] Abkürzungen: ATC = Aspartat-Transkarbamylase; ATP = Adenosintriphosphat; CTP = Zytidintriphosphat; UTP = Uridintriphosphat; 6-AUR = 6-Azauridin

erfolgte durch Homogenisation der entsprechenden Gewebe in KCl-Lösung (0,14 M KCl und 2,5 mM NaOH) und Zentrifugation bei 15000 *g* (30 min). Der erhaltene Überstand wurde für die Enzymbestimmung verwendet. Die Inkubation erfolgte in Triäthanolamin-Puffer (0,1 M, *p*H 9,2). Als Bezugsstandard diente eine wässrige Lösung von Ureidobernsteinsäure (10^{-3} M) der Firma Calbiochem. Die zum Inkubationsmedium zugesetzten Metabolite wurden käuflich von der Firma Reanal bezogen. 6-Aza-Uridin (6-AUR) war ein Produkt der Firma Spofa, United Pharmaceutical Works, Praha.

Die Aktivität der ATC wird in der vorliegenden Arbeit in nMol/mg Std. angegeben und stellt somit die spezifische Aktivität dar. Die Proteinbestimmung erfolgte nach der Biuret-Methode. Die in den Tabellen angegebenen Werte wurden aus mindestens 5 Einzelversuchen ermittelt. Die Aktivitätsangaben stellen somit Mittelwerte ± Standardabweichungen dar.

Ergebnisse

Der Einfluß von zwei verschiedenen ATP-Konzentrationen auf die Aktivität der ATC aus normaler Leber, Hepatomen und WALKER-Karzinomen ist in der Tabelle 1 dargestellt. Aus dieser Tabelle geht deutlich hervor, daß die Aktivität der ATC aus normaler Leber kleiner ist als aus Hepatomen und WALKER-Karzinomen. Allerdings konnten wir in unseren Versuchen die

Tabelle 1

Der Einfluß von ATP auf die ATC-Aktivität verschiedener Gewebe

Gewebe	ATP [M]	Aktivität [$nMol \cdot mg^{-1} \cdot Std.^{-1}$]	Veränderung der Aktivität [%]
normale Leber	–	224 ± 31,8	–
normale Leber	10^{-2}	120 ± 40,0	−47,0
normale Leber	10^{-3}	149 ± 22,0	−33,5
Hepatom	–	308 ± 19,6	–
Hepatom	10^{-2}	222 + 18,6	−28,0
Hepatom	10^{-3}	328 ± 16,0	+ 6,5
WALKER-Karzinom	–	570 ± 41,5	–
WALKER-Karzinom	10^{-2}	403 ± 34,8	−29,4
WALKER-Karzinom	10^{-3}	610 ± 19,0	+ 7,0

von CALVA und COHEN [2] beobachtete Erhöhung der Enzymaktivität in Hepatomen nicht völlig erreichen. Diese Tatsache ist vermutlich auf eine gewisse Vermischung der Hepatome mit normalem Lebergewebe zurückzuführen, da nur eine makroskopische Trennung beider Gewebearten erfolgen konnte. Dagegen ist die Aktivität der ATC aus WALKER-Karzinomen gegenüber normaler Leber um über 100% erhöht, wodurch der von CALVA und COHEN [2] angedeutete Zusammenhang zwischen der Aktivität dieses Enzyms und der Wachstumspotenz bestätigt wird. Auch die Beeinflußbarkeit der ATC-Aktivität durch ATP ist in den untersuchten Geweben sehr unterschiedlich und deutet auf eine Durchbrechung der normalen Regulierbarkeit dieses für die Nukleotidsynthese notwendigen Enzyms in den unkontrolliert wachsenden Hepatomen und WALKER-Karzinomen. Die relativ hohe ATP-Konzentration von 10^{-2} M ruft in allen untersuchten Geweben eine, wenn auch stark unterschiedliche Hemmung des Enzyms hervor. Dagegen ist es auffällig, daß eine ATP-Konzentration von 10^{-3} M in normalem Lebergewebe noch eine beträchtliche Hemmung der ATC-Aktivität verursacht, während das Enzym aus Hepatomen und WALKER-Karzinomen durch diese ATP-Konzentration nicht mehr gehemmt wird, sondern sogar eine geringe Steigerung seiner Aktivität erfährt. Wenn diese unterschiedlichen Hemmungen auf einer echten Beeinflussung der ATC-Aktivität beruhen, könnten sie eine wesentliche Bedeutung für die Steuerung des normalen Wachstums und für die relative Unabhängigkeit des bösartigen Wachstums besitzen.

CTP ist ein normales Endprodukt der Synthesekette, deren erste Stoffwechselreaktion durch die ATC katalysiert wird. Auf Grund der erwähnten Untersuchungen von GERHART und PARDEE [1] war es von Interesse, den Einfluß des CTP auch auf die Aktivität der ATC aus Säugetiergeweben zu untersuchen. Die Ergebnisse unserer Versuche an den genannten tierischen Geweben brachten den überraschenden Befund, daß CTP in diesen Geweben überhaupt keinen Einfluß auf die Aktivität der ATC besitzt (Tab. 2). Die geringfügige Steigerung der Enzymaktivität aus Hepatomen durch CTP ist nicht signifikant, so daß auch für dieses Gewebe ein regulatorischer Einfluß auf die ATC verneint werden muß. Anhand der fehlenden Regulationswirkung dieses Metaboliten kann also keine Differenzierung zwischen Geweben, die einer normalen Wachstumskontrolle unterliegen und den unkontrolliert wachsenden Tumoren vorgenommen werden.

Tabelle 2

Der Einfluß von CTP auf die ATC-Aktivität verschiedener Gewebe

Gewebe	CTP [M]	Aktivität [nMol · mg^{-1} · Std.$^{-1}$]	Veränderung der Aktivität [%]
normale Leber	–	224 ± 31,8	–
normale Leber	10^{-3}	224 ± 28,6	± 0,0
Hepatom	–	308 ± 19,6	–
Hepatom	10^{-3}	355 ± 24,0	+11,5
WALKER-Karzinom	–	570 ± 41,5	–
WALKER-Karzinom	10^{-3}	568 ± 39,8	± 0,0

Da neben CTP auch UTP ein Endprodukt der gleichen Stoffwechselfolge ist, könnte man auch von diesem Metaboliten eine Feedback-Regulationswirkung auf die ATC erwarten. Die Ergebnisse entsprechender Untersuchungen sind in der Tabelle 3 dargestellt. Danach unterliegt das Enzym aus normaler Leber einer starken Aktivitätsregulierung durch UTP, während eine Feedback-Hemmung in WALKER-Karzinomen nicht gefunden werden konnte. Eine gegenüber normaler Leber erniedrigte Aktivitätsbeeinflussung

Tabelle 3

Der Einfluß von UTP auf die ATC-Aktivität verschiedener Gewebe

Gewebe	UTP [M]	Aktivität [nMol · mg^{-1} · Std.$^{-1}$]	Veränderung der Aktivität [%]
normale Leber	—	224 ± 31,8	—
normale Leber	10^{-3}	134 ± 30,7	−40,0
Hepatom	—	308 ± 19,6	—
Hepatom	10^{-3}	254 ± 25,0	−17,6
WALKER-Karzinom	—	570 ± 41,5	—
WALKER-Karzinom	10^{-3}	555 ± 31,3	− 2,5

ergab sich für die ATC der Hepatome. Diese geringere Feedback-Inhibition ist möglicherweise ein Ausdruck einer im Vergleich zu WALKER-Karzinomen verminderten Wachstumspotenz und der wesentlich geringeren Malignität der Hepatome.

Durch die Untersuchungen von REYES und HEIDELBERGER [7] konnte festgestellt werden, daß die Kinasen, die Nukleoside in Nukleotide überführen, relativ unspezifisch sind und deshalb auch Substanzen mit geringfügigen Strukturabwandlungen gegenüber den natürlichen Metaboliten umsetzen können. Somit wäre zu erwarten, daß auch 6-AUR bzw. seine phosphorylierten Derivate ebenso wie das UTP einen Einfluß auf die ATC-Aktivität ausüben könnten, der dann als „Pseudofeedback-Regulation" zu bezeichnen wäre. Aus diesem Grunde war es von Interesse, den Einfluß von 6-AUR auf die Aktivität der ATC in verschiedenen Geweben zu untersuchen und mögliche Unterschiede in dieser „Pseudofeedback-Regulation" zwischen den einzelnen Geweben zu erkennen. Die Ergebnisse dieser Untersuchungen sind in den Tabellen 4 und 5 wiedergegeben. Aus der Tabelle 4, in der der Einfluß verschiedener 6-AUR-Konzentrationen auf die Aktivität der ATC aus normaler Leber, Hepatomen und WALKER-Karzinomen dargestellt ist, geht hervor, daß durch diese Substanz die Aktivität der ATC aus Leber mit über 30% eine beträchtliche Hemmung erfährt, während der Einfluß auf kanzerisierte Gewebe deutlich geringer ist.

In der Tabelle 5 ist der kombinierte Einfluß von 6-AUR und ATP auf die Aktivität der ATC wiedergegeben. Auf Grund der bereits erwähnten Untersuchungen von REYES und HEIDELBERGER [7] muß angenommen werden, daß 6-AUR mittels ATP in Aza-Uridinmono-, -di- oder -triphosphat überführt wird, so daß ein direkter Vergleich mit der Wirkung des UTP möglich wird. Im Gegensatz zu der fehlenden Beeinflußbarkeit der ATC aus WALKER-Tumoren durch UTP führt der kombinierte Einfluß von 6-AUR und ATP noch zu einer merklichen Hemmung der ATC-Aktivität. Außerdem wird auch deutlich, daß die hemmende Wirkung des Gemisches 6-AUR + ATP viel stärker ist, als die Summe der Wirkungen beider Substanzen. Dieser Einfluß ist jedoch bei WALKER-Tumoren viel stärker ausgeprägt als bei der

Tabelle 4

Der Einfluß von 6-Azauridin auf die ATC-Aktivität verschiedener Gewebe

Gewebe	6-Azauridin [M]	Aktivität [nMol · mg^{-1} · Std.$^{-1}$]	Veränderung der Aktivität [%]
normale Leber	—	224 ± 31,8	—
normale Leber	$5 \cdot 10^{-2}$	139 ± 20,0	−37,7
normale Leber	$5 \cdot 10^{-3}$	155 ± 25,8	−31,0
Hepatom	—	308 ± 19,6	—
Hepatom	$5 \cdot 10^{-2}$	267 ± 10,8	−13,5
Hepatom	$5 \cdot 10^{-3}$	349 ± 18,6	+13,0
WALKER-Karzinom	—	570 ± 41,5	—
WALKER-Karzinom	$5 \cdot 10^{-2}$	409 ± 35,5	−28,2
WALKER-Karzinom	$5 \cdot 10^{-3}$	523 ± 40,0	− 8,0

Tabelle 5

Der Einfluß von 6-Azauridin + ATP auf die ATC-Aktivität verschiedener Gewebe[1]

Gewebe	6-Azauridin + ATP [M]	Aktivität [nMol · mg^{-1} · Std.$^{-1}$]	Veränderung der Aktivität [%]
normale Leber	—	224 ± 31,8	—
normale Leber	$5 \cdot 10^{-2}$	43 ± 18,0	−80,8
normale Leber	$5 \cdot 10^{-3}$	138 ± 21,6	−38,8
Hepatom	—	308 ± 19,6	—
Hepatom	$5 \cdot 10^{-2}$	103 ± 18,4	−66,6
Hepatom	$5 \cdot 10^{-3}$	310 ± 21,3	± 0,0
WALKER-Karzinom	—	570 ± 41,5	—
WALKER-Karzinom	$5 \cdot 10^{-2}$	38 ± 13,2	−93,4
WALKER-Karzinom	$5 \cdot 10^{-3}$	434 ± 43,0	−22,4

[1] Die ATP-Konzentrationen betragen jeweils 20% der angegebenen Azauridin-Konzentrationen

normalen Leber, während die Hepatome wieder eine Mittelstellung einzunehmen scheinen. Aus der Beeinflußbarkeit der ATC durch 6-AUR bzw. 6-AUR + ATP ist zu schließen, daß die allosterischen Zentren dieses Enzyms die im 6-AUR vorliegende abgewandelte Struktur eines normalen Metaboliten offensichtlich nicht „erkennen" können. Daraus resultiert eine gewisse Unspezifität der ATC, die möglicherweise für die allgemeine Toxizität des Antimetaboliten 6-AUR mit verantwortlich sein könnte.

Diskussion

Die vorliegenden Ergebnisse zeigen, daß sich die Regulierbarkeit der ATC durch die normalen Metabolite ATP und UTP in den unkontrolliert wachsenden Tumoren anders verhält als im normalen Lebergewebe. Sofern sich diese Ergebnisse, die auf Grund von in vitro-Untersuchungen erhalten wurden, auch auf die lebende Zelle übertragen lassen, können sie möglicherweise in das allgemeine Erscheinungsbild der „Wachstumsregulation" bösartiger Tumore einbezogen werden. Bemerkenswert ist die Tatsache, daß die Wirkung der normalen Metabolite auf die ATC-Aktivität von Hepatomen und WALKER-Karzinomen stets in einer Verminderung der im normalen Lebergewebe zu beobachtenden Hemmung der Enzymaktivität besteht und niemals eine erhöhte Regulationskapazität zeigt. Wie bereits in der Einleitung angedeutet wurde, spielt die Ansprechbarkeit von Schlüsselenzymen auf regulierende Metabolite für die Regulation der gesamten Stoffwechselprozesse eine wesentliche Rolle. Die vorliegenden Ergebnisse lassen den Schluß zu, daß die Akzeptoren für den Angriff der regulierenden Metabolite im Molekül der ATC von Hepatomen oder WALKER-Karzinomen entweder eine geringere Affinität für den „allosterischen" Metaboliten besitzen oder überhaupt nicht ausgebildet sind. Derartige Veränderungen allosterischer Zentren bzw. die Aufhebung der normalen Feedback-Regulation in Hepatomen ist bereits von SIPERSTEIN und FAGAN [8] für die Cholesterin-Synthese beschrieben worden. Sie ist, ebenso wie die Veränderungen der Feedback-Regulation der ATC in Hepatomen und WALKER-Karzinomen, eine der möglichen Ursachen für die Aufhebung der normalen Stoffwechselregulationen und damit für den Prozeß der Differenzierung in Tumoren.

Das in der Tabelle dargestellte Fehlen jeglicher Beeinflußbarkeit der ATC durch CTP steht im klaren Gegensatz zu der von GERHART und PARDEE [1] entdeckten Feedback-Regulation der ATC durch diesen Metaboliten in E. coli. Ein weiterer Gegensatz ergibt sich aus dem unterschiedlichen Einfluß des ATP auf die ATC aus tierischen Geweben und aus E. coli. ATP induziert in diesen Mikroorganismen eine schwache Steigerung der Enzymaktivität. Allerdings besteht dieser Unterschied, wie die Tabelle 1 zeigt, nur für hohe ATP-Konzentrationen und bei ATP-Konzentrationen von 10^{-3} M sogar nur für das normale Lebergewebe. Die Regulationsverhältnisse bei ATP-Konzentrationen von 10^{-3} M in Hepatomen und WALKER-Karzinomen erinnern dagegen an die entsprechenden Verhältnisse in Bakterien, so daß sich die Frage nach einem möglichen Zusammenhang zwischen den Regulationsmechanismen der ATC und dem Grad der Ausbildung einer spezifischen Zelleistung, wie sie von den untersuchten Geweben nur die normale Leber besitzt, ergeben. Jedoch umfaßt gerade diese Frage ein sehr komplexes Geschehen, so daß weitere Untersuchungen zur Abklärung dieser Problematik notwendig erscheinen. Die Einbeziehung abgewandelter Metabolite in diese Untersuchungen führte zu dem Ergebnis, daß die Wirkung von Antimetaboliten nicht auf bösartige Tumoren beschränkt bleibt, sondern auch normale, nicht wachsende Gewebe in fast gleicher Weise betrifft. Diese Tatsache äußert sich in der hohen Toxizität des 6-AUR und findet durch die vorliegenden Untersuchungen eine gewisse biochemische Ergänzung.

Literatur

[1] GERHART, J. C. u. A. B. PARDEE: Cold Spring Harbor Sympos. quantitat. Biol. **28**, 491 (1963).
[2] CALVA, F. u. P. P. COHEN: Cancer Res. **19**, 101, 679 (1959).
[3] KIM, S. u. P. P. COHEN: Arch. Biochem. Biophysics **109**, 421 (1965).
[4] HEISE, E. u. M. GÖRLICH: Exp. Cell Res. **33**, 289 (1964).
[5] KORITZ, S. B. u. P. P. COHEN: J. biol. Chemistry **209**, 145 (1954).
[6] SPEKTOR, L., M. E. JONES u. F. LIPMANN in: Methods in Enzymology. Hrg.: S. P. COLOWICK u. N. O. KAPLAN. New York: Acad. Press. **3**, 653 (1957).
[7] REYES, PH. u. CH. HEIDELBERGER: Mol. Pharmacol. **1**, 14 (1965).
[8] SIPERSTEIN, M. D. u. V. M. FAGAN: Advances in Enzyme Regulation **2**, 249 (1964).

Summary

E. HEISE and M. GÖRLICH: Feedback regulation of the aspartate transcarbamylase from different tissues

The influence of adenosine triphosphate, cytidine triphosphate, uridine triphosphate and 6-aza-uridine on the activity of aspartate transcarbamylase was studied in the normal liver, hepatoma and WALKER's carcinomas and differences in the regulability of this enzyme from normal and cancerous tissues noticed. The effect of the above substances on aspartate transcarbamylase activity is always lower in homogenates from cancerous tissues than in normal liver homogenates. The aspartate transcarbamylase is additionally subjected to a "pseudofeedback regulation" by 6-aza-uridine which is present in any tissue investigated, and is regarded as a biochemical expression of a relatively unspecific effect of this antimetabolite.

Резюме

Е. Гейзе и М. Гёрлих: Регуляция обратной связи (feedback regulation) аспартатранскарбамилазы из разных тканей

Изучали влияние аденозинтрифосфата, цитидинтрифосфата, уридинтрифосфата и 6-аза-уридина на активность аспартаттранскарбамилазы в нормальной печени, гепатомах и карциномах Уокера, причем обнаруживались различия в регулируемости этого энзима из нормальных и раковых тканей. Влияние указанных веществ всегда меньше в гомогенатах из раковых тканей, чем в нормальных печеночных гомогенатах. Аспартаттранскарбамилаза дополнительно подвергнута так называемой „pseudo-feedback regulation" благодаря 6-аза-уридину, которая распространяется на все исследованные ткани и является биохимическим выражением довольно неспецифичного эффекта этого антиметаболита.

Acta biol. med. german., Band 17, Seite 25—33 (1966)

Aus dem Forschungsinstitut für Tuberkulose und Lungenkrankheiten Berlin-Buch
(Direktor: Prof. Dr. med. habil. P. STEINBRÜCK)

Lungengewebe-, Herzblut- und Urinspiegel nach Inhalation und Injektion von Penizillin und Streptomyzin

G. PICKROTH

(Eingegangen am 25. 8. 1965)[1]

Zusammenfassung

Mit Ultraschall lassen sich Aerosole hoher Nebeldichte bei lungengängigen Partikelgrößen herstellen. Um einen Einblick zu gewinnen, mit welchen Wirkungen bei der Inhalation von Ultraschall-Nebel zu rechnen ist, wurden bei Meerschweinchen Inhalationen mit Penizillin- und Streptomyzin-Nebeln durchgeführt und die Lungengewebe-, Blut- und Urinspiegel bestimmt. Die gleichen Testungen erfolgten nach i. m. Injektion und Instillation der Antibiotika in den Darmtrakt. Schon nach relativ kurzer Inhalationszeit sind hohe Antibiotikakonzentrationen im Atemtrakt feststellbar. Biologisch wirksame Antibiotikaspiegel lassen sich in den untersuchten Materialien selbst mehrere Tage nach einer einmaligen Inhalation noch nachweisen. Der resorptive Abtransport der Antibiotika nach Inhalation erfolgt vorwiegend über die Pulmonalvene. Insgesamt geht aus den Untersuchungsergebnissen hervor, daß nach Inhalation von Aerosolen, die mit Ultraschall erzeugt sind, im Atemtrakt nachhaltige Wirkungen zu erwarten sind. Sie sind für die in den vorliegenden Untersuchungen verwendeten Antibiotika bei Applikation gleicher Dosen nach Inhalation höher einzuschätzen als nach der Injektion.

Mit Hilfe der Ultraschallzerstäubung gelingt es, Nebeldichten bis über 300 mg Flüssigkeit/l Luft bei Tröpfchengrößen zu erreichen, die im Durchmesser sämtlich unter 5 μ liegen. Damit sind die entscheidenden Voraussetzungen gegeben, um schon mit kurzen Inhalationen von Aerosolen im Atemtrakt erhebliche Konzentrationen an gelösten Substanzen und damit entsprechende Wirkungen zu erzielen. Die vorliegenden tierexperimentellen Untersuchungen sollen dazu dienen, nähere Einblicke in diese Fragen zu gewinnen und helfen, Probleme des Abtransportes inhalierter Substanzen aus dem Atemtrakt zu studieren. Von Interesse ist dabei ausschließlich der Abtransport auf resorptivem Wege.

Die Lösung der Aufgabe wird dadurch erschwert, daß bei der Inhalation von Aerosolen Nebelanteile im Mund und im Rachen vorzeitig sedimentieren und in den Magen–Darm-Kanal abwandern. Deshalb können für die vorgesehenen Untersuchungen nur solche Substanzen Verwendung finden, die im Magen–Darm-Kanal zerstört bzw. nicht resorbiert werden. Weitere Schwie-

[1] Nach Revision am 24. 2. 1966.

rigkeiten liegen darin, daß im Tierexperiment bei der Inhalation von Aerosolen eine ganze Reihe nicht übersehbarer Faktoren, auf die z. T. noch eingegangen wird, gegeben sind. Aus diesem Grund können die vorliegenden Untersuchungen nur orientierenden Charakter haben.
Um die gestellten Fragen zu beantworten, sollen die Lungengewebe–Herzblutserum- und Urinspiegel nach Inhalation oder i.m. Injektion einer Testsubstanz verglichen und dabei auf die applizierten Mengen Bezug genommen werden. Mit den gleichen Untersuchungen nach Instillation des Teststoffes in den Magen–Darmtrakt ist dann festzustellen, ob die Ergebnisse bei Inhalation der Mittel durch Resorption aller geschluckter Teststoffanteile aus dem Darmtrakt beeinflußt werden.

Methodik

Auf Einzelheiten der Methodik kann im Rahmen dieser Veröffentlichung nicht eingegangen, sondern nur eine grobe Übersicht gegeben werden.
Als Testsubstanz für die Durchführung der Untersuchungen erwiesen sich auf Grund umfangreicher Vorversuche die Antibiotika Penizillin *G* und Streptomyzinsulfat, als Versuchstiere Meerschweinchen, gut geeignet. Insgesamt fanden 245 Tiere Verwendung.
Die Untersuchungen erfolgten in Gruppen von je 4 Tieren an jeweils verschiedenen Tagen.

Applikation der Antibiotika

Inhalation: Zur Herstellung der Ultraschall-Aerosole sowie zu den technisch-physikalischen Kenndaten der verwendeten Nebel wird auf bereits erfolgte Veröffentlichungen [2] hingewiesen. Hier sei nur vermerkt, daß die Vernebelfrequenz 2,7 MHz betrug. Die Applikation des Penizillin- und Streptomyzin–Ultraschall-Aerosols erfolgte in einem Kleinraumgefäß, in dem gerade 4 Meerschweinchen Platz hatten. Die Tiere hockten eng aneinander. Häufig versuchten sie ihre Köpfe untereinander zu verstecken oder schoben ihre Nasen in das Fell eines anderen Tieres. Dadurch wurde das Einatmen des Aerosols offensichtlich behindert. Eine sichere Aussage zu der während einer bestimmten Zeit tatsächlich inhalierten Aerosolmenge zu machen, ist nicht möglich. So wechseln z. B., um nur das Wichtigste zu nennen, bei Bewegung der Tiere die Nebeldichten, durch die Nasenatmung sedimentieren erhebliche Aerosolanteile bereits in den Nasengängen, auf dem Fell der Tiere setzen sich Aerosolpartikel ab, die Nebel müssen als Raumaerosole appliziert werden, die sich durch Koagulation rasch in ihren technisch-physikalischen Größen verändern. Bisher gibt es keine Methode die Wiederausatmungsrate des Aerosols in einem solchen Tierversuch sicher zu erfassen.
Ein Anhalt zur Beurteilung der Ergebnisse ist dadurch zu gewinnen, daß die Gesamtmenge an Penizillin oder Streptomyzin, die als Aerosol in das Kleinraumgefäß gelangt, der inhalierten Menge gleichgesetzt wird. Der Fehler, mit dem die eingeatmete Aerosolmenge dadurch um ein Vielfaches als zu groß angenommen wird, erlaubt, den jeweils ermittelten Wert als Mindestwert anzusehen. Für jedes Meerschwinchen betrug die in das Kleinraumgefäß als Aerosol zugeführte Penizillinmenge während der Inhalationszeit von 45 min 70000 IE, die Streptomyzinmenge für die gleiche Zeit 130 mg.
Injektion: Penizillin wurde von Tiergruppe zu Tiergruppe ansteigend i.m. injiziert, d. h. bis zu einer Menge, bei welcher der Lungenparenchymspiegel ähnlich hoch lag wie nach der Inhalation. Diese Dosen betrugen 30000, 70000 und 200000 IE Penizillin.
Für Streptomyzin lag die Injektionsdosis bei 250 mg, entsprechend dem DAB 6. Nachtrag, British pharmacope = 200000 IE Streptomyzin. Diese Dosis war für die Meer-

schweinchen toxisch, insgesamt aber doch noch so verträglich, daß die Tiere, außer einigen, bis zum Abtöten am Leben blieben. Mit dieser eben noch verträglichen Dosis sollten hohe Lungengewebespiegel erzielt werden.
Instillation: Je Meerschweinchen wurden 200000 IE Penizillin oder 250 mg Streptomyzin in den Magen instilliert. Bei diesen Tieren erfolgten die Gewebe-, Blutserum- und Urinspiegeluntersuchungen wie nach Inhalation und Injektion der Antibiotika. Gleichzeitig wurde der Magen-Darminhalt sowie das Darmgewebe getestet.

Materialentnahme und Verarbeitung des Materials

Unmittelbar nach Antibiotikaapplikation wurden die Tiere unter fließendem Wasser gewaschen, der Brustkorb in Äthernarkose eröffnet und Herzblut durch Punktion entnommen. Die Meerschweinchen wurden durch Ausbluten getötet. Danach erfolgte die Entnahme der Lunge. Urin wurde nach Spalten der Bauchdecken durch Blasenpunktion gewonnen. Eine Urinentnahme war bei einigen Meerschweinchen nicht möglich, da diese Tiere mit Beginn der Narkose spontan entleerten. Der Tracheobronchialbaum (Tbr) wurde aus dem Lungenparenchym (Lgp) soweit wie möglich herauspräpariert, meist rissen zuletzt die feinen Bronchialäste. Um Antibiotikaanteile nicht aus der Lunge auszuschwemmen, wurde auf die Durchspülung des Blutgefäßsystems des kleinen Kreislaufs verzichtet, so daß auch das im Blut der Lungengefäße befindliche Antibiotikum in die Antibiotikaspiegel des Atemtraktgewebes mit eingeht. Es wird den tatsächlichen Antibiotikawert im Gewebe nach Inhalation niedriger, nach Injektion höher erscheinen lassen.
Da ein ständiger Abtransport der inhalierten Antibiotika aus der Lunge stattfindet, ist auch die Zeit bis zum Beginn der Materialverarbeitung von Bedeutung. Während bei den ersten Tieren vom Ende der Antibiotikainhalation bis zur abgeschlossenen Entnahme des Materials eine Zeit bis zu 19 min verging, verkürzten sich die Zeiten mit steigender Übung rasch auf 6 min. Die Materialentnahme bei den Injektionsgruppen erfolgte 15 min nach Injektion, da entsprechend den Vorversuchen zu diesem Zeitpunkt die höchsten Werte im Lgp zu erwarten waren. Die Materialentnahmen nach Instillation wurden 15, 30, 45 und 60 min nach der Applikation vorgenommen, um die jeweils höchsten Werte für die einzelnen Gewebe, für das Herzblutserum und Urin zu ermitteln.
Von besonderem Interesse im Hinblick auf die Resorption war das Verhalten der Antibiotikaspiegel im Verlauf von Tagen nach einer einmaligen Inhalation oder Injektion. Die Spiegel wurden bis zu 6 Tagen nach einer einmaligen Applikation der Antibiotika getestet.
Die Aufbereitung des Lgp und Tbr erfolgte durch Feinmörserung mit sterilem, gewaschenem und getrocknetem Seesand. Anschließend wurde dem Gewebebrei eine Phosphatpufferlösung von jeweils 1 ml pro g Gewebe zugegeben und die Mischung nach sorgfältiger Schüttelung bei +4 °C für 1 Std. im Kühlschrank aufbewahrt. Als Testmaterial diente jeweils 1 ml des überstehenden Zentrifugates, 1 ml Serum des Blutes der rechten sowie der linken Herzkammer und jeweils 1 ml Urin.
Die Testungen der Antibiotikakonzentrationen erfolgten gegen den Staphylokokkusstamm S.G. 511[1] sowie Sarcina lutea PCI 1001 ATCC. Dazu fand ein Pepton-Agar Verwendung, in dem Testsporen des Staphylokokkusstammes S.G. 511 oder Sarcina lutea PCT 1001 ATCC eingemischt waren. Der Pepton-Agar wurde erwärmt in Spezialschalen eingegossen. Nach Erstarren wurden mit einem Rundstanzer Löcher zur Aufnahme der Antibiotika-Standardlösungen (Vergleichslösungen) und des Testmaterials im Agar ausgehoben.
Der jeweils wiedergegebene, aus der Verdünnung berechnete Antibiotikawert stellt einen Durchschnittswert von 4 Testungen des entsprechenden Materials dar. Die

[1] Herrn Dr. GREULICH, VEB Jenapharm, sind wir für die Unterstützung der Untersuchungen zu besonderem Dank verpflichtet

Werte sind in IE (= IE/g oder IE/ml) und μg (= μg/g Gewebe oder μg/ml) wiedergegeben.
Die *rechnerisch-statistische Auswertung* der Ergebnisse erfolgt in Anlehnung an Weber [3].

Ergebnisse

Folgend werden nur die wichtigsten Ergebnisse der Untersuchungen mitgeteilt.

Inhalation

Penizillin: Nach der Inhalation des Penizillinaerosols finden sich im Lgp der einzelnen Tiere etwas unterschiedliche Antibiotikamengen. Sie schwanken zwischen 15,6 IE und 110 IE mit einem Mittel von 45,6 IE, bei Einzelwerten am häufigsten um 40 IE. Im Tbr ergeben sich Penizillinmengen zwischen 8,8 und 160 IE mit einem Durchschnittswert von 64,4 IE. Das Ergebnis von 8,8 IE fällt deutlich aus dem Gesamtrahmen heraus. Im Blutserum der rechten Herzkammer lassen sich Mengen zwischen 0,7 und 10,5 IE mit einem Mittel von 4,61 IE feststellen. Im Blutserum der linken Herzkammer finden sich die Werte zwischen 1,1 und 20,0 IE und einem Durchschnitt von 7,97 IE deutlich höher als rechts. Die Mittelwerte im Urin liegen bei 250 IE. Der Konzentrationsunterschied an Penizillin zwischen Tbr (64,4 IE) und Lgp (45,6 IE) ist für den Tbr statistisch nicht signifikant größer, also wahrscheinlich nur zufällig bedingt. Die Differenz der Werte zwischen Lgp (45,6 IE) zum Herzblutserum links (7,97 IE) und rechts (4,61 IE) ist statistisch signifikant.
Das besondere Augenmerk erfordert das Verhältnis des Antibiotikaspiegels des Blutserums der linken zur rechten Herzkammer. Die errechnete Prüfgröße p beträgt 3,5. Nach Weber [3] (Tabelle 5, Seite 420) ist die Differenz der arithmetischen Mittel mit p kleiner als 1% und damit statistisch signifikant. Das bedeutet aber, daß das Ergebnis des höheren Blutserumspiegelwertes in der linken Herzkammer nicht zufällig zustande kommt, sondern der Ausdruck eines grundsätzlichen Verhaltens ist.
24 Std. nach der Inhalation von 70000 IE Penizillin (Abb. 1) ist der Lgp-Spiegel mit 0,08 IE im Mittel stark abgesunken. Die Urinspiegel finden sich mit 15 IE in einem guten therapeutischen Bereich. Wenn der 28-Std.-Wert im Lgp etwas höher als der 24-Std.-Wert liegt, dann ist durchaus die Frage zu stellen, ob bei den Untersuchungen nach 24 Std. unter Umständen ein methodischer Fehler vorliegt, der sich aber nicht mehr sichern ließ. Am 4. Tage sind bei einem Tier noch 0,16 IE Penizillin in der Lunge nachweisbar, bei einem anderen 0,2 IE im Urin. Die höchsten Konzentrationen der im gleichen Zeitrhythmus ermittelten Urinspiegel entsprechen dabei dem jeweils höchsten Lgp.
Die Differenz zwischen dem Herzblutserum links (52,3 IE) und rechts (48,4 IE) ist gering und nicht als statistisch signifikant zu sichern. Statisti-

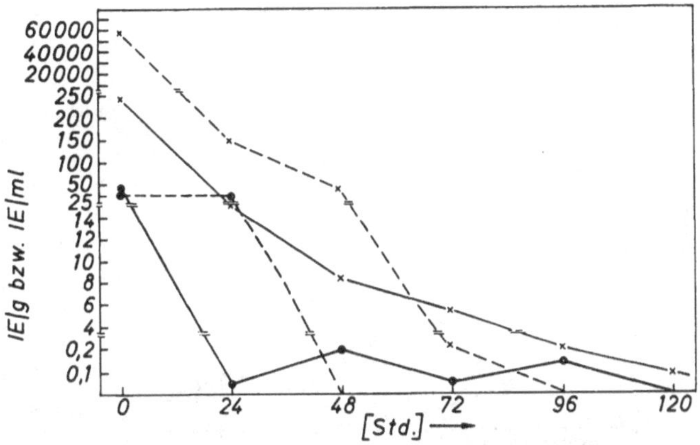

Abb. 1. Graphische Darstellung der Lungengewebe- und Urinspiegel für Penzillin nach einmaliger Inhalation oder Injektion.

Lungengewebespiegel in IE/g: ·———· Inhalation; · – – – – · Injektion. Urinspiegel in IE/ml: x———x Inhalation; x – – – x Injektion. Abszisse: Zeit nach Applikation

sche Signifikanz der Werte des Blutserums der rechten und linken Herzkammer zur Penizillinkonzentration des Lgp (10,3 IE) und des Tbr besteht. Die Differenz zwischen Tbr (14,3 IE) und Lgp (10,3 IE) sowie Herzblutserum links (52,36 IE) und rechts (48,43 IE) ist statistisch nicht signifikant.

Streptomyzin: Nach der Inhalation findet sich im Lgp ein Mittelwert von 50,43 μg, im Tbr von 65,9 μg. Bei Schwankungen der Einzelergebnisse der Testungen des Antibiotikums zwischen 11 und 48 μg bzw. zwischen 6 und 23 μg beträgt das arithmetische Mittel für das Blutserum der linken Herzkammer 11,9 μg. Im Urin finden sich durchschnittlich Streptomyzinmengen von 40 μg.
Die Mittelwerte im Tbr, Lgp, im Herzblut rechts und links verhalten sich in ihrer Relation ähnlich wie nach Penizillininhalation. Die Streuung der Einzelwerte um ihr arithmetisches Mittel ist am höchsten für die Antibiotikakonzentration im Tbr. Eine bessere Vorstellung vermittelt aber der Variabilitätskoeffizient, da er die Streuung in von 100 des jeweiligen arithmetischen Mittels ausdrückt. Er liegt für den Tbr und das Lgp bei über 67%, für das Herzblutserum rechts sogar über 77,8%.

Als wichtigstes Ergebnis aber ergibt sich aus den statistischen Untersuchungen, daß der Wert des Serums der linken Herzkammer für Streptomyzin ebenso wie nach Penizillininhalation statistisch signifikant höher liegt als rechts.

20 Std. (Abb. 2) nach der Inhalation beträgt die Antibiotikamenge im Lgp im Durchschnitt 18,4 und am sechsten Tage noch 2,6 μg. Die Werte für den Urin liegen bei 25 bzw. 1,2 μg.

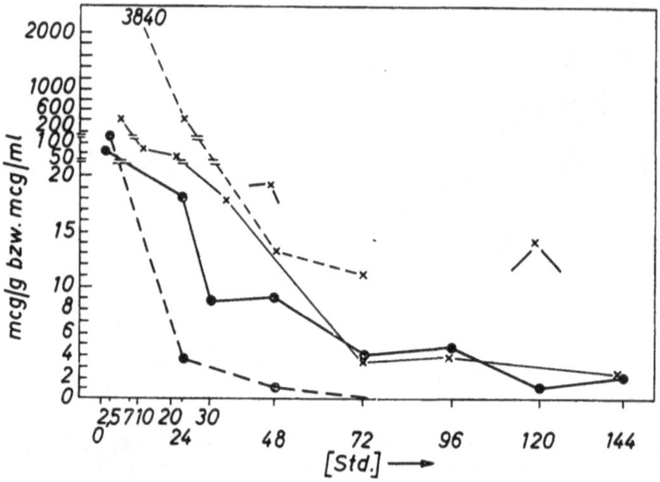

Abb. 2. Graphische Darstellung der Lungengewebe- und Urinspiegel für Streptomyzin nach einmaliger Inhalation oder Injektion.
Lungengewebespiegel in mcg/g: •——• Inhalation; • – – – • Injektion.
Urinspiegel in mcg/ml: x——x Inhalation; x – · – x Injektion

Injektion

Penizillin: Nach der i.m. Injektion variieren die Antibiotikaspiegel entsprechend der applizierten Menge von Tiergruppe zu Tiergruppe erheblich. Beträgt der Mittelwert für das Lgp nach 30000 IE i.m. 1,58 IE, so liegt er nach 20000 IE i.m. bei 39,5 IE. Für das Blutserum links ergeben sich nach Injektion von insgesamt 200000 IE im Durchschnitt 135,6 für das Herzblutserum rechts 121,8 IE.

Die Differenz zwischen dem Herzblut links (52,3 IE) und rechts (48,4) ist gering und statistisch nicht signifikant. Statistische Signifikanz von Blut der rechten und linken Herzkammer zur Penizillinkonzentration des Lgp (10,3 IE) und Tbr besteht. Die Differenz zwischen Tbr (14,3 IE) und Lgp (10,3 IE) sowie Herzblutserum links (52,36 IE) und Serum rechts (48,43 IE) ist nicht statistisch signifikant zu sichern.

24 Std. (Abb. 1) nach Injektion von 200000 IE finden sich im Lgp durchschnittlich 21,5 IE, nach 48 Std. ist das Antibiotikum nicht mehr nachweisbar. Die Urinwerte sinken von 82 IE nach 24 Std. auf 0,21 IE am 3. Tag ab. Am 4. Tag ist das Antibiotikum nicht mehr festzustellen.

Streptomycin: Die Injektionsdosis ist aus eingangs erwähnten Gründen mit 250 mg Streptomyzin pro Tier sehr hoch gewählt. Dementsprechend ergibt die Antibiotikabestimmung Durchschnittswerte, die für das Lgp bei 131,7 µg das Herzblutserum links und rechts bei 1314 bzw. 1370 µg erheblich sind. Die Verhältniszahlen lassen erkennen, daß die Lgp-Konzentrationen etwa ein Zehntel der jeweiligen Blutkonzentrationen an Streptomyzin betragen. Der Durchschnittswert an Streptomyzin im Blutserum der linken Herzkammer liegt um ein Geringes höher als rechts. 24 Std. (Abb. 2) später finden sich im Durchschnitt im Lgp nur noch 3,8 µg, der Urinwert beträgt 250 µg. Nach 48 Std. ist Streptomyzin im Lgp nicht mehr nachzuweisen, im Urin noch 11 µg.

Instillation

Es sind hier nur die jeweils höchsten, nach Instillation ermittelten Werte wiedergegeben. Für Penizillin fanden sich nach Instillation im Lgp Werte von 1 IE, für Streptomyzin von 0,8 µg. Im Blut konnte Streptomyzin mit maximal 8,5 µg links 9 IE nachweisbar. Nach 60 min fanden sich im Serum nachgewiesen werden. Für Penizillin waren im Blutserum rechts 11 IE, rechts und links nur noch je 0,1 IE Penizillin.

Diskussion

Bereits nach kurzzeitigen Inhalationen von Penizillin und Streptomyzin als Ultraschall-Aerosol lassen sich beim Meerschweinchen hohe Konzentrationen der Antibiotika in der Lunge, im Blut und Urin nachweisen. Die einzigen in der Literatur ermittelten, in ähnlicher Weise beim Kaninchen durchgeführten Inhalationsversuche mit Streptomyzin als Düsenaerosol zeigen wesentlich geringere Werte. Nach einer 70 min andauernden Streptomyzininhalation (keine Angaben der technisch-physikalischen Daten zum Gerät und zum Aerosol) und erst dann, wenn mit dem Antibiotikum Hyaluronidase appliziert wurde, fand sich ein Lungengewebespiegel von 8 µg und ein Blutspiegel von 2 µg Streptomyzin (Lass [1]).
Aus den Untersuchungsergebnissen geht hervor, daß bei Inhalation allein der Atemtrakt Eintrittspforte für das Antibiotikum in den Organismus ist. Obgleich bei den vorliegenden Untersuchungen das Lungenparenchym-, Tracheobronchialbaum-, Blutserum- und Urinspiegel von Tier zu Tier teilweise sogar erheblich schwanken, ergeben sich letztlich beim Einzeltier immer wieder ähnliche Relationen. Das betrifft insbesondere das Verhältnis der Werte der Lunge zum Blutserum sowie des Blutserums der Herzkammer links zu rechts. Sie weisen darauf hin, daß die Pulmonalvene der wichtigste Abtransportweg der Antibiotika ist. Ein praktisch umgekehrtes Verhalten der Blutserum- und Lungenparenchym-Konzentrationen zeigt sich nach Injektion der Antibiotika. Für die Streptomyzinwerte nach Injektion sind wegen der verwendeten toxischen Dosen die Untersuchungsergebnisse nur mit Einschränkung zu verwerten.
Trotz ähnlicher Ergebnisse jeweils nach Inhalation oder Injektion sind gewisse Unterschiede zwischen den Streptomyzin- und Penizillinwerten nicht zu übersehen. Während sich die Lungenkonzentration nach Inhalation der Antibiotika gleichen, liegen die Blutspiegel nach Streptomyzinhalationen im Vergleich zu den Lungenspiegeln deutlich höher. Im Hinblick auf das Molekulargewicht wurde eine schlechtere Resorption erwartet. Allerdings sagt das Molekulargewicht nichts über die nicht bekannte Molekülgröße aus. Umgekehrt zeigt sich die Differenz der Werte vom Blut zum Lungengewebe nach

Injektion wesentlich größer als beim Penizillin. Die Frage, ob ein Vergleich der Lungengewebe-, Blut- und Urinspiegel im Hinblick auf die unterschiedliche Dosierung möglich ist, muß noch kurz besprochen werden. Die bei der Inhalation im Tierversuch tatsächlich inkorporierte Antibiotikamenge exakt zu bestimmen, ist mit den bisherigen Methoden nicht möglich. Vielleicht helfen hier Untersuchungen mit radioaktiv markierten Aerosolen weiter. Die Basis zum Vergleich bietet sich, wie eingangs erwähnt, darin an, daß die pro Tier in Aerosolform zugeführte Antibiotikamenge der inkorporierten Menge gleichgesetz wird. Daß nur ein Bruchteil davon in den Atemtrakt kommt, wird nochmals betont. Der bewußt in Kauf genommene grobe Fehler zu ungunsten der Inhalation kommt aber letztlich der Einschätzung der Inhalation von Aerosolen insofern entgegen, als dadurch die ermittelten Werte aussagekräftig werden.

Insgesamt wird mit den vorliegenden Untersuchungen festgestellt, daß durch die Inhalation von Aerosolen im Atemtrakt hohe Konzentrationen an Wirksubstanzen erreicht werden können. Damit aber sind die Voraussetzungen einer nachhaltigen lokalen Wirkung gegeben.

Die Blutserumuntersuchungen zeigen darüber hinaus, daß sich unter den gegebenen Bedingungen selbst eine inhalative Allgemeinbehandlung durchführen läßt.

Werden die Antibiotikaspiegel nach einer einmaligen Inhalation über längere Zeit in den Lungen und im Urin verfolgt, so ergibt sich, daß Streptomyzin bis zu 5 Tagen eine gute, Penizillin bis zu 4 Tagen eine ausreichend biologische Wirksamkeit zeigt. Offensichtlich gelangt ein Teil des inhalierten Antibiotikums rasch, ein anderer langsam ins Blut.

Nach der Injektion großer Dosen Penizillin und Streptomyzin kommt es zwar zu hohen Ausgangskonzentrationen in der Lunge, die offensichtlich von dem im Blut des kleinen Kreislaufs befindlichen Antibiotikums mitbestimmt werden. Sie fallen wahrscheinlich auch deshalb in kurzer Zeit rasch ab. Werden die Konzentrationen der Antibiotika in Lunge, Blut und Urin als maß der Wirksamkeit angesehen, so zeigen die Untersuchungsergebnisse, daß die Inhalation von Streptomyzin- und Penizillinaerosolen eine wirksame Applikationsmethode ist. Sie können zunächst nur auf das gesunde Meerschweinchen bezogen werden. Inwieweit die Ergebnisse auf den Menschen übertragbar sind, muß zunächst offen bleiben. Unabhängig davon aber zeichnen sich hier wichtige und bedeutungsvolle Aufgaben ab, deren Klärung das Ziel späterer Untersuchungen sein muß.

Literatur

[1] Lass, A.: Die resorptionssteigernde Wirkung der Hyaluronidase bei der Inhalation tuberkulostatischer Substanzen. Beitr. Klin. Tuberkul. **106**, 104 (1951).

[2] Pickroth, G.: Ultraschall- und Düsenaerosole in der Medizin. VEB Gustav-Fischer-Verlag, Jena 1963.

[3] Weber, E.: Grundriß der biologischen Statistik. 2. Aufl. VEB Gustav-Fischer-Verlag, Jena 1956.

Anschrift des Verfassers: Dr. med. habil. G. Pickroth, Forschungsinstitut für Tuberkulose und Lungenkrankheiten Berlin-Buch, Karower Str. 11.

Summary

G. Pickroth: Levels of penicillin and streptomycin in lung tissue, cardiac blood, and urine, as administered by inhalation and injection

Finely atomized aerosols can be produced by ultrasonic procedure for medical purposes. The effect of its inhalation has been tested on guinea pigs exposed to penicillin and streptomycin, and its levels in the lung tissue, blood and urine subsequently determined. Similar tests were made after intramuscular injection and instillation of antibiotics into the intestinal tract. High concentrations of antibiotics are detectable in the respiratory tract already after a relatively short time of inhalation. Biologically active amounts of antibiotics in the investigated material can be demonstrated even several days after a single inhalation. Its resorptive removal is predominantly by the pulmonary vein. The results indicate that lasting effects may be expected in the respiratory tract from the inhalation of aerosols produced by ultrasonic procedure. For the antibiotics used in this study at equal dosages, the effect is higher with inhalation than with injection.

Резюме

Г. Пикрот: Ингаляция и инъекция пинициллина и стрептомицина с последующим определением их уровня в легочной ткани, сердечной крови и в моче

Методом ультразвука создаются аэрозоли высокой плотности с величиной частиц, свободно проходящих через легкие. Для проверки ожидаемых эффектов при ингаляции ультразвукового тумана, морских свинок подвергали ингаляции пенициллинным и стрептомицинным туманом с последующим определением их уровня в легочной ткани, крови и моче. Те же самые определения проводились после внутримышечной инъекции и инстилляции антибиотиков в кишечный тракт. Уже после сравнительно короткой ингаляции обнаруживаются высокие концентрации антибиотиков в дыхательном тракте. Биологически действенные уровни антибиотиков среди исследованных веществ наблюдаются даже спустя несколько дней после разовой ингаляции. Резорптивный транспорт ингалированных антибиотиков проходит главным образом по легочной вене. Полученные данные показывают, что при ингаляции аэрозолей, созданных ультразвуковым методом, достигается длительный эффект в дыхательном тракте. Для использованных в данных экспериментах антибиотиков можно ожидать большего эффекта от ингаляции, чем от инъекции.

Acta biol. med. german., Band 17, Seite 34—43 (1966)

Laboratoire de Biochimie Centre Expérimental „23 August" — Bucarest

Valeurs biochimiques sériques et syndrome de suprasollicitation chez le sportif

G. Haralambie

(Reçu le: 1. 11. 1965)

Résumé

On a déterminé chez les sportifs, le matin, au repos et à jeun, les valeurs sériques des protéines totales, des fractions électrophorétiques, des glyco- et lipoprotéines, de l'acide sialique et de l'urée.
Les sujets en régime d'effort difficile présentent — lorsque les résultats sportifs sont très bons — un tableau sérique très proche de la normale. Au contraire, chez les sujets systématiquement suprasollicités, et dont les résultats sportifs sont mauvais, on trouve un tableau caractéristique: hypoprotéinémie avec rapport A/G bas, des valeurs modérémment augmentées des glycoprotéines et nettement accrues pour l'acide sialique, l'urée et les alpha-2 globulines. La connaissance de l'ensemble de ces modifications est importante pour la définition objective de l'état biologique du sportif de performance.

L'entraînement sportif a pour but l'adaptation graduelle des sujets à l'effort considérable requis par les hautes performances modernes. Il s'agit d'un processus complèxe, même en ne tenant compte que du côté purement biologique. En effet, les multiples variables — nature de l'effort, intensité, durée, volume total de travail fourni, rythmicité pause-travail, conditions du milieu — influencent l'organisme à divers degrès au point du vue quantitatif, mais aussi qualitatif. Lorsque ce qu'on appelle «le dosage de l'effort» est mal choisi, les résultats sportifs recherchés ne sont point atteints et un excés de travail conduit en fin de compte à l'état de surentraînement. Mais entre les deux extrêmes: bonne préparation, suivie de succès sportifs et surentraînement — état pathologique — il y a certainement des gradations, des situations intermédiaires, réversibles si on arrive à les mettre en évidence à temps, et par conséquent, à cette dernière condition, encore du domaine de la physiologie et non de la pathologie de l'effort.
Des indices biochimiques d'une sollicitation physique exagérée ont été décrits assez récemment par les auteurs français et japonais. Chailley-Bert et al. [3], [4], Plas et al. [12] ont trouvé, après quelques jours d'effort intense, une élévation du niveau de l'urée et de l'acide urique du sérum, avec excrétion accrue de 17-cétostéroides.
Les observations de Sumiyoshi [16] chez la femme, se rapportent chez de sujets suivant un programme d'entraînement dur, à une diminution de

l'hémoglobine, du potassium et du rapport albumines/globulines dans le sérum. YOSHIMURA et al. [18], [19] trouvent, après une semaine d'entraînement dur au rugby, une diminution de l'hémoglobine ainsi que des protéines totales du sang. Nous avons noté, chez des sportifs de diverses spécialités, en programme d'entraînement intense, une augmentation notable des glycoprotéines sériques, de l'acide sialique et de l'urée [6], [7], avec parfois des valeurs plutôt basses des protéines totales. Enfin des auteurs finlandais trouvent aussi une diminution de l'hémoglobine lors des sollicitations sportives importantes et durables [13].

Les aspects mentionnés se rapportent non point à l'influence immédiate de l'exercice sportif, mais à l'action cumulée des efforts succéssifs et sont retrouvés même après 12 à 20 heures de repos.

En tenant compte des faits mentionnés, nous avons cherché à établir les principaux caractères biochimiques de ce que l'on peut désigner comme «l'état de suprasollicitation» chez le sportif. Cet état doit être distingué, dans notre opinion, de celui de surentraînement, parce que facilement reversible après la modification des paramètres de l'effort, et par une médication vitaminique appropriée; il ne peut être considéré, comme le surentraînement, un état pathologique et se rencontre plus souvent que ce dernier dans la pratique sportive.

Sujets et méthodes

La caractérisation, au point de vue métabolique, d'un état dévié de la normale, suppose la possibilité de comparaison avec les données ,,normales", établies au milieu d'une même population, par les mêmes procédés de dosage, sur un nombre acceptable de sujets. Nous avons choisi à cette fin les résultats des analyses effectuées sur des sportifs roumains de haute classe, sains, au repos depuis 36 à 48 heures, le matin et à jeun (cette dernière situation est valable pour tous les examens exposés dans ce travail). Les valeurs normales des glycoprotéines sont prises chez des sujets soumis à un examen clinique sévère, en excluant les cas avec tonsillite chronique, antécédents hépatiques ou granulomes dentaires; tous ces sujets avaient d'ailleurs la valeur relative des γ-globulines entre 14 et 19%.

Les procédés de dosage utilisés sont mentionnés dans nos travaux antérieurs [6], [8]. Les résultats se trouvent dans le tableau 1.

Un second groupe de sportifs est constitué par des sujets qui ont eu à une date rapprochée (6 à 8 jours au plus) de notre examen, des succès sportifs notables. Il comprend 4 champions nationaux, deux composants des équipes nationales, maîtres ès sports, et trois ,,seconde place". Ici encore, pas d'indices pathologiques et un bon état subjectif (tab. 2).

Le troisième groupe comprend les sujets que nous avons considéré comme étant suprasollicités au moment de l'examen; ils avaient un programme de travail très chargé, accusaient subjectivement un état de fatigue assez accentuée et leur rendement sportif était nettement inférieur à celui réalisé quelques semaines auparavant ou après. Chez plusieurs, les examens médico-sportifs habituels donnaient aussi des indices de désadaptation[1].

[1] Nous remercions les drs. AL. PARTHENIU, T. MINESCU, I. WIESENMAIER et ECAT. CRISTEA pour les observations en ce sens qu'ils nous ont communiqué

Résultats et discussion

Les valeurs trouvées chez les sportifs considérés normaux et en dehors de l'activité physique, correspondent à ceux que nous avons établi antérieurement [6] avec cependant des chiffres plus élevés pour les protéines totales; elles sont superposables aux données normales des hommes sains, non sportifs [5], [10].

Tableau 1

Valeurs biochimiques sériques chez des sportifs sains, au repos depuis 36 à 48 heures

Facteur étudié	No. cas	Moyenne ± D.S.	Valeurs extrêmes
Protéines totales [g%]	30	6,95±0,33	6,38 – 7,77
Glycoprotéines (mg% hexoses)	30	121,7 ±7,49	105 – 134
mg hexoses/g protéines	30	17,5 ±1,17	15,3 – 20
Acide sialique [mg%]	19	58 ±5,55	50 – 65
Electrophorèse	18		
Protéines totales [g%]		7,08±0,25	6,60 – 7,57
Albumines %		59,1 ±1,98	56 – 62,2
Globulines %			
α_1		5,7 ±0,96	3,8 – 7,5
α_2		6,8 ±1,1	5,3 – 9,4
β		10,2 ±1,09	9,3 – 12,2
γ		17,5 ±0,96	15,9 – 19,3
Rapport A/G		1,45±0,13	1,27 – 1,65
Albumines [g%]		4,18±0,23	3,80 – 4,55

Une situation semblable se retrouve chez les sujets du tableau 2, quoique les investigations ont été faites dans une période d'activité sportire intense (avant ou après un concours). Une certaine tendance à l'augmentation de l'urée sérique peut être notée chez ces sujets, fait explicable par le régime d'effort suivi [3], [12]. Les autres facteurs du métabolisme protidique restent cependant dans les limites normales, ce qui permet de formuler l'hypothèse que, chez les sportifs très bien entraînés, en bon équilibre physiologique, les perturbations des facteurs biochimiques étudiés, provoquées par l'effort, ne sont probablement pas profondes et certainement pas durables.

Par contre, chez les sportifs considérés comme en état de suprasollicitation (tabl. 3), les valeurs des divers facteurs diffèrent significativement de celles des deux autres groupes de sportifs. L'ensemble des valeurs paraît constituer un tableau caractéristique, où se retrouvent à la fois les aspects partiels signalés par divers auteurs [3], [4], [12], [16], [18], [19] à côté d'autres, chez des sportifs soumis à un travail fatigant, répété.

Les valeurs basses des protéines sériques — chez 10 sujets de 12 — sont dues à une diminution des albumines, en valeur relative et absolue. L'urée dépasse le chiffre normal maximum de la méthode (400 mg/litre) chez 11 des 12 sujets. On trouve une élévation très nette des glycoprotéines sériques,

Tableau 2

Sujets ayant un très bon rendement sportif; analyses effectuées peu avant ou après le concours, le matin, au repos et à jeun

Sport	Protéines sériques [g%]	Urée [mg/l]	Glyco-protéines [mg%]	mg hexoses / g protéines	Acide sialique [mg%]	mg ac. sial. / g protéines	Rapport A/G	Albu-mines [g%]	α_2-globu-lines [%]
1 Pentathlon	6,83	355	118,5	17,3	52,5	7,7	1,65	4,25	7,1
2 Pentathlon	7,57	380	132	17,4	61,5	8,1	1,50	4,55	6,5
3 Athlètisme	7,27	350	113	15,5	54	7,4	1,41	4,21	7,2
4 Basketball	6,99	530	133	19	56	8	1,54	4,24	7,4
5 Pentathlon	7,29	365	138	19	65	8,9	1,23	4,02	8,1
6 Athlètisme	6,97	415	126,5	18,1	63	9	1,33	3,98	7
7 Athlètisme	6,70	370	126	18,8	58	8,6	1,38	3,89	5,7
8 Handball	7,78	440	125	16	58	7,4	1,34	4,45	5,8
9 Athlètisme	7,00	325	140	20	64	9,1	1,62	4,33	5,5
$M_{\bar{x}} \pm$ D S.	7,17 ± 0,41	381 ±62,4	128±9	17,9±1,48	59,1±4,5	8,35±0,69	1,44 ±0,13	4,21 ±0,22	6,7 ±0,87

Note le sujet no. 4 a èté examiné le lendemain du dernier match du championnat national (une rencontre difficile chaque soir, durant 5 jours).

Tableau 3

Sportifs considérés en état de suprasollicitation physique, avec résultats mauvais précédant notre examen

Sport	Protéines sériques [g%]	Urée [mg/l]	Glyco-protéines [mg%]	mg hexoses / g protéines	Acide sialique [mg%]	mg ac. sial. / g protéines	Rapport A/G	Albu-mines [g%]	α_2-globu-lines [%]
1 Sky	6,40	825	132	20,6	65	10,1	0,90	2,96	10,3
2 Sky[1]	5,70	645	136	23,8	82,5	14,4	1,02	2,88	9,9
3 Sky[1]	6,38	520	142	22,5	75	11,7	0,92	3,06	10,2
4 Sky[1]	6,57	510	133	20,2	72	10,9	0,90	3,09	10,3
5 Hòckey[1]	6,48	660	136,5	21	83	12,9	1,18	3,51	8,3
6 Tennis	6,16	380	130,5	21	80	13	1,07	3,19	8,2
7 Aviron	6,25	610	139	22,2	78,5	12,5	1,13	3,32	8,1
8 Aviron	6,30	440	142,5	22,6	72	11,4	1	3,15	12
9 Athlètisme	6,99	425	153	21,9	77,5	11,1	1,12	3,70	7,5
10 Athlètisme	6,40	515	146	22,8	75	11,7	1,24	3,57	9,7
11 Basketball	7,12	880	159	22,3	80	11,2	1,10	3,73	11,8
12 Haltères	6,38	530	138,5	21,7	75	11,7	0,88	3	8,2
$M_{\bar{x}} \pm$ D S	6,43 ± 0,37	580±154	141 ±8,56	21,8±1,03	76±5,12	11,9±1,15	1,04 ±1,10	3,26 ±0,30	9,5 ±1,48
Signification de là différence vis-à-vis du tableau 2: p	<0,001	<0,005	<0,005	<0,001	<0,001	<0,001	<0,001	<0,001	<0,001

[1] Note les sujets marqués d'un astérisque présentaient une réaction au thymol (McLAGAN) positive (entre 4,5 et 6,7 unités)

plus accentuée pour leur composante acide sialique que pour celle des hexoses. Le rapport: mg hexoses/gr protides ne dépasse pas 24, les chiffres trouvés chez des sujets portant une inféction focale ou autre maladie inféctieuse étant de l'ordre de 24 à 28. Les γ-globulines sont d'ailleurs entre des limites admissibles, sans dépasser 20%. L'élévation de l'acide sialique et des α_2-globulines est semblable à celle rencontrée dans les états inflammatoires subaïgus.

Les modifications signalées sont sans doute étroitement liées à une dynamique endocrine spéficique, que nous n'avons pas étudiée; diverses observations ont d'ailleurs montré que les sujets fortement sollicités manifestent un état hypercatabolique protidique, avec augmentation de l'élimination des 17-hydroxycorticoïdes et diminution du rapport 17-hydroxy-/17-désoxycorticoïdes [2], [11].

Continuer à pratiquer des entraînements intenses, répétés, dans un tel état biologique c'est sans doute risquer α' arriver en fin de compte à des troubles de la santé. Par contre, nous avons des indications qu'en diminuant le volume et l'intensité du travail, en augmentant les intervalles de repos et en modifiant l'alimentation et la vitaminisation des sportifs, on réalise assez rapidement un rétablissement du rendement sportif, de l'état subjectif et du tableau sérique des sujets.

Les tests utilisés peuvent donc permettre au médecin de se rendre compte en temps utile de l'effet d'une sollicitation physique exagérée et par conséquent d'intervenir pour empêcher l'installation de l'état de surentraînement.

Afin de vérifier la concordance entre les données de laboratoire et la situation réelle, nous avons entrepris des dosages, dans les mêmes conditions, chez 5 autres sujets, dont l'activité dans la période antérieure à l'examen était de nature à les amener dans une situation pouvant être rapprochée de celle des sujets du tableau 3. Les résultats se trouvent au tableau 4 et confirment l'existence d'un ensemble de valeurs biochimiques caractéristiques à l'état de suprasollicitation physique.

Au point de vue théorique, deux aspects méritent une certaine attention:

a) Les sujets suprasollicités présentent une augmentation des glycoprotéines sériques, nettement plus accentuée en ce qui concerne l'acide sialique que pour les hexoses. En effet, le rapport hexoses/acide sialique est pour les «normaux» de l'ordre de 2 à 2,2 pour les sujets du tableau 2 il dépasse toujours 2, tandis que chez les «suprasollicités» il est en moyenne de 1,80 (un seul cas: 2,12). On a supposé que la fraction glycoprotéique riche en hexoses augmente lors des procéssus de «réparation» après une agression [10], [14], [17]. Au contraire, l'augmentation des fractions riches en acide sialique traduirait une lésion biochimique au niveau du tissu mésenchymateux, à partir duquel elles seraient mises en liberté. L'augmentation relative et absolue de la fraction α_2 globulinique est en effet compatible avec une réaction inflammatorie et destruction modérée de tissus [15]; cette augmentation est assez évidente chez nos sujets suprasollicités.

Tableau 4

Valeurs sériques des sportifs considérés en état de suprasollicitation (vérification des conclusions du tableau 3)

Sport	Protéines totales [g%]	Urée [mg/l]	Glyco-protéines [mg%]	mg hexoses / g proteides	Acide sialique [mg%]	mg ac. sialique / g protéines	Rapport A/G	Albu-mines [g%]	α_2-globu-lines [%]
Handball	6,40	510	144,5	22,5	68	10,6	1,22	3,52	9,2
Athlètisme	6,40	440	132,5	20,7	80	12,5	1,27	3,58	9,1
Athlètisme	6,38	530	138,5	21,7	75	11,7	0,98	3,16	10,5
Boxe	6,43	560	132	20,5	75	11,5	1,28	3,60	9
Boxe	6,46	430	131,5	20,3	67	10,4	1,35	3,72	8,6
$M_{\bar{x}} \pm$ D.S.	6,41 ±0,03	494 ±56,8	136,5 ±5,68	21,1 ±0,93	73 ±5,3	11,3 ±0,86	1,22 ±0,14	3,51 ±0,21	9,3 ±0,72

Il s'agirait donc, chez le sportif sollicité systématiquement au delà de sa capacité d'adaptation momentanée, de la prédominance des procéssus de dégradation sur ceux de réparation; ce fait a été discuté par BUGARD [2] en partant d'observations faites sur d'autres facteurs biochimiques. Le niveau élevé de l'urée sérique indique d'ailleurs aussi un catabolisme protidique accentué.

Cet «état de souffrance» du mésenchyme se traduit finalement, sur le plan sportif, par l'augmentation de l'incidence des accidents musculotendineux; nous avons déjà signalé le niveau élevé de l'acide sialique chez les sportifs qui ont présenté bientôt après de tels accidents [7]. Et le fait est confirmé par nos données plus récentes.

b) La prédominance des procéssus de dégradation sur ceux de l'anabolisme est soulignée par la diminution de la fraction albuminique sérique chez les suprasollicités. Nous avons constaté que l'effort physique important mais non épuisant conduit à l'augmentation tardive des protéines totales ainsi que du rapport A/G, en confirmant une autre observation due à JOHNSON et WONG [9]. Il reste à préciser si la diminution du rapport A/G chez les sujets des tableaux 3 et 4 est due à une augmentation de la dégradation des albumines ou à une diminution de leur synthèse hépatique. On a montré, chez le singe, que le rapport A/G diminue nettement après hypophysectomie, pour se rétablir consécuti vement à l'administration de hydrocortisone [1]; on revient ainsi aux observations de BUGARD [2], mentionnées plus haut.

S'il s'agit réellement d'un déséquilibre entre la synthèse et la dégradation protidique — comme les faits le font supposer — les repercussions de ce phénomène pourraient se faire sentir au niveau du muscle et des organes; ceci serait compatible avec la perte de poids, toujours signalée chez les sujets surentraînés.

Conclusion

Des recherches effectuées en appliquant notre méthodologie chez environ 400 sportifs adultes, en partie publiées [6]- [8], nous ont permis d'esquisser la caractérisation de plusieurs types de déviations du tableau protidique sérique normal. Le type «suprasollicitation» discuté dans ce travail, se différencie assez nettement des autres déviations métaboliques qu'on rencontre plus souvent chez le sportif; le diagnostic est facilité par la connaissance du rendement sportif et de l'activité antérieure du sujet; les accuses subjectives ne font que le confirmer. Nous considérons comme aspects caractéristiques du syndrôme de suprasollicitation chez le sportif: l'urée sérique nettement augmentée ainsi que l'acide sialique, les hexoses liés aux protéines de l'ordre de 130—160 mg% (normal: 100—130 mg%; états inféctieux: 145 à 230 mg% et γ-globulines augmentées, ce que nous ne trouvons pas lors de la suprasollicitation), rapport mg hexoses/g protides entre 20—23, augmentation relative et absolue de la fraction α_2-globulinique et diminution relative et absolue des albumines. Le rapport A/G se situe entre 0,9 et 1,3 et les protéines totales sont généralement diminuées (au-dessous de 6,5 g%).
La connaissance de l'ensemble de ces éléments est importante pour le physiologiste et le médecin sportif sur le plan théorique et pratique: elle permet de préciser en même temps que d'approfondir une situation physiologique qui se rencontre plus souvent que nos données ne le permettraient de le supposer chez le sujet pratiquant l'entraînement sportif moderne. Nos observations ne font que souligner la nécessité des examens de laboratoire complexes chez les sportifs.

Adresse de l'auteur: Dr. G. Haralambie, Centre Expérimental „23 August" Laboratoire de Biochimie, Bd. Muncii 37—39, Bucarest/Roumanie.

Bibliographie

[1] Beardwood, C., C. Gilbert u. J. Gilman: Endocrine control of the plasma A/G ratio. Nature [London] **195**, 710 (1962).
[2] Bugard, P.: La fatigue. Paris: Masson 1960.
[3] Chailley-Bert, P., F. Plas u. G. Pallardy: Fondements biologiques d'une alimentation rationelle adaptée au sportif. XIII. Intern. Kongreß f. Sportmed., Wien. Sportärztl. Prax. **3**, 89 (1960).
[4] Chailley-Bert, P., F. Plas u. G. Pallardy: Le métabolisme protidique au cours de l'effort prolongé. Presse méd. **70**, 705 (1962).
[5] Eastham, R. D.: Valori biochimici in Medicina. Roma: Il pensiero Scientifico 1961.
[6] Haralambie, G.: La valeur de certaines constantes biochimiques du sérum chez les sportifs en régime d'entraînement intense. Acta biol. med. german. **13**, 30 (1964).

[7] HARALAMBIE, G. u. G. JEFLEA: Indices biochimiques du sérum et la récupération après l'effort physique. Int. Z. angew. Physiol. **20**, 515 (1965).
[8] HARALAMBIE, G.: Les modifications tardives des protéines sériques dans l'effort physique. Fiziol. norm. si pat. **11**, 85 (1965).
[9] JOHNSON, T. u. H. WONG: Effects of training and competitive swimming on serum proteins. J. appl. Physiol. **16**, 807 (1961).
[10] KELLEN, J.: Die Eiweißzucker. Leipzig: Thieme 1960.
[11] MILCOU, St. u. I. STEFAN in: Medicina culturii fizice. Sous la réd. de Fl. C. ULMEANU. Bucarest: Éd, Médicale 1965, S. 93.
[12] PLAS, F., P. BUGARD u. G. PALLARDY: Le métabolisme protidique au cours de l'effort prolongé. Méd., Educ. physique Sport. **35**, 215 (1961).
[13] ROMPOTTI, K.: Blood tests as a guide training. In: Fred WILT: Run, run, run. Los Altos: Track and Field News Inc. 1964, S. 186.
[14] SHETLAR, M.: Serum glycoproteins. Ann. New York Acad. Sci. **94**, 44 (1961).
[15] SCHUMACHER, G. u. H. SCHLUMBERGER: Über Veränderungen der alpha$_2$-Globuline des Serums. Dtsch. med. Wschr. **88**, 645 (1963).
[16] SUMIYOSHI, K.: A study on the changes of body fluid and the circulatory system of athletes. Abstr. of Papers pres. at the Intern. Congress of Sport Sci. Tokyo, 1964, S. 66.
[17] WHITE, N., M. SHETLAR u. J. SCHILLING: Ann. New York Acad. Sci. **94**, 297 (1961).
[18] YOSHIMURA, H., T. YAMADA, Sh. KIMURA u. O. AIKO: On the mechanism and physiological role of anemia caused by strenuous training to sports. Abstr. of Papeıs pres. at the Intern. Congress of Sport Sci. Tokyo, 1964, S. 127.
[19] YOSHIMURA, H., T. YOSHIOKA, T. YAMADA u. J. MOMOTA: Studies on the prevention of anemia due to hard physical training. Abstr. of Papers pres. at the Intern. Congress of Sport Sci. Tokyo, 1964, S. 128.

Zusammenfassung

G. HARALAMBIE: Biochemische Serumwerte und Überlastungssyndrom bei Leistungssportlern

Bei Leistungssportlern wurden morgens nüchtern in Ruhe die Serumwerte für Gesamtprotein, Eiweißzucker, Lipoproteide, Sialinsäure und Harnstoff sowie die elektrophoretischen Fraktionen bestimmt.

Sportler, im schweren Trainingsprogramm mit guten Leistungen, wiesen Serumwerte auf, die denen ruhender Sportler weitgehend entsprachen. Bei überbelasteten Sportlern mit schlechten Wettkampfleistungen treten charakteristische Veränderungen auf und zwar Hypoproteinämie mit erniedrigtem Albumin-Globulinquotienten, leicht erhöhte Werte für Eiweißhexosen und hohe Werte für Sialinsäure, α_2-Globuline und Harnstoff. Die Veränderungen werden als objektive Kriterien für die Feststellung des biologischen Zustandes von Hochleistungssportlern gewertet.

Summary

G. HARALAMBIE: Biochemical serum levels and the syndrome of overstrain in sportsmen

The serum levels of total protein, glycoprotein and lipoprotein, sialic acid, urea as well as the electrophoretic fractions of highly active sportsmen were determined in the morning, the stomach empty, at rest.

Sportsmen under conditions of severe physical activity, achieving good results, revealed serum levels which widely agreed with those of resting sportsmen. Overstrained sportsmen attaining poor success, showed characteristic changes, such as hypoproteinemia with reduced levels of the albumin-globulin ratio, slightly enhanced levels of protein hexoses and high values of sialic acid, α_2-globulins and urea. The changes are regarded as objective criteria for evaluating the biologic condition of sportsmen.

Резюме

Г. Гараламби: Биохимические параметры сыворотки и синдром перегруженности у спортсменов

У спортсменов определились, утром натощак в состоянии покоя, сывороточные уровни общего протеина, гликопротеина, липопротеидов, сиаловой кислоты и мочевины, а также электрофоретические фракции. Спортсмены, выполняющие тяжелую тренировочную программу и достигающие хороших успехов, обнаруживали величины, хорошо согласующиеся с величинами отдыхающих спортсменов. У перегруженных спортсменов с плохими успехами в состязаниях наблюдались характерные изменения, как гипопротеинемия с пониженным отношением альбуминов к глобулинам, слегка повышенным уровнем белковых гексоз и высоким уровнем сиаловой кислоты, α_2-глобулинов и мочевины. Наблюдаемые изменения рассматриваются как объективные критерии для оценки биологического состояния спортсменов.

Acta biol. med. german., Band 17, Seite 44–47 (1966)

Aus der I. Medizinischen Universitätsklinik Szeged, Ungarn
(Direktor: Prof. Dr. M. JULESZ)

Die Wirkung des Hexadimethrin-bromids auf den Askorbinsäuregehalt der Nebennieren der Ratte

K. KOVÁCS und M. A. DÁVID

(Eingegangen am 15. 2. 1966)

Zusammenfassung

Hexadimethrin-bromid (5 mg i.v.) bzw. Verbindung 48/80 (0,1 mg i.p.) rufen bei Ratten Askorbinsäuredepletion in den Nebennieren hervor. Eine Vorbehandlung der Tiere mit Substanz 48/80 verhindert die Hexadimethrin-bromid-bedingte Askorbinsäureverminderung. Phenergan läßt sie indessen unbeeinflußt. Die Befunde sprechen dafür, daß die Herabsetzung des Askorbinsäuregehaltes durch Hexadimethrin auf indirektem Wege über eine Histaminliberation erfolgt.

Histaminzufuhr aktiviert bei Ratten das Hypophysen-Nebennierenrinden-System und verursacht in den Nebennieren eine Askorbinsäuredepletion [3], [7], [12], [14], [15]. Eine ähnliche Wirkung entfaltet nach NASMYTH [8] auch der Histamin-Liberator „Substanz 48/80".

Im Laufe unserer Untersuchungen erhob sich die Frage, ob eine adrenokortikale Askorbinsäureverminderung auch nach Hexadimethrin-bromid zustandekommt. Es verursacht wie Substanz 48/80 Mastzelldisruption und Histaminmobilisierung [4], [5]. Außerdem entfaltet es einen beträchtlichen Einfluß auf die histologische Struktur und die Hormonproduktion der Nebennierenrinde. Hexadimethrin-bromid bewirkt bei Ratten ausgedehnte Nekrosen in der Zona glomerulosa bei rund der Hälfte der Tiere Herdinfarkte in der Zona fasciculata und reticularis, sowie eine bedeutende Herabsetzung der Hormonproduktion [1], [6]. Unsere Versuche zeigten, daß 1 Std. nach Hexadimethrin-bromid-Injektion – d. h. vor dem Zustandekommen der Nebennierennekrose – der Askorbinsäuregehalt der Nebennieren weitgehend verringert war.

Es schien daher lohnend zu untersuchen, ob diese Depletion durch Vorbehandlung mit dem Histaminliberator 48/80, bzw. durch Antihistaminverabreichung zu verhindern ist. Die vorliegende Arbeit berichtet über diesbezügliche Ergebnisse.

Methodik

Versuchstiere waren 180–220 g schwere, bei Standarddiät gehaltene, weibliche Albinoratten des gleichen Stammes, die in sechs Gruppen unterteilt wurden:

1. Gruppe. Kontrollen. 1 Std. vor Tötung 0,5 ml physiologische Kochsalzlösung i.v.

2. Gruppe. 1 Std. vor Tötung 5 mg (0,5 ml) Hexamethrin-bromid i.v. (Polybrene, Abbott).

3. Gruppe. 1 Std. vor Tötung 0,1 mg (0,1 ml) Substanz 48/80 i.p.

4. Gruppe. Entleerung der Histamindepots nach FELDBERG und TALESNIK [2]. 1. Tag 2×0,1; 2. Tag 2×0,2; 3. Tag 2×0,3; 4. Tag 2×0,4; 5. Tag 2×0,5 mg Substanz 48/80 i.p. Am 6. Tag etwa 12 Std. nach letzter Injektion Tötung.

5. Gruppe. 48/80-Vorbehandlung wie Gruppe 4. 1 Std. vor Tötung 5 mg Hexadimethrin-bromid i.v.

6. Gruppe. 3 Std. vor Tötung 5 mg Phenergan (Specia, Paris) s.c. und ¼ Std. vor Tötung 5 mg Phenergan i.p., 1 Std. vor Tötung 5 mg Hexadimethrin-bromid.

Die Tiere wurden durch Dekapitation getötet, unmittelbar darauf die linke Nebenniere herauspräpariert, gereinigt, ihr Gewicht an der Torsionswaage auf 0,5 mg genau bestimmt und ihr Askorbinsäuregehalt mit der ROE-KUETHERschen 2,4-Dinitrophenylhydrazin-Methode [10] im HAVEMANN-Photometer ermittelt. Die Askorbinsäuremengen sind in mg/100 g Nebennierengewicht ausgedrückt.
Die Ergebnisse wurden biometrisch mit dem Student „t"-Test bewertet.

Ergebnisse

Die Ergebnisse sind tabellarisch zusammengefaßt. Nach Hexadimethrinbromid war der Askorbinsäuregehalt der Nebennieren im Verhältnis zu dem der Kontrollen signifikant verringert. Eine beträchtliche Askorbinsäuredepletion verursachte auch die einmalige i.p. Injektion der Substanz 48/80. Dies steht im Einklang mit den Befunden von NASMYTH [8]. Die fünftägige Vorbehandlung mit Substanz 48/80 dagegen hatte keine nennenswerte Veränderung im Askorbinsäuregehalt der Nebennieren zur Folge. Eine wesentliche Verminderung kam im Falle der Hexadimethrin-bromid-Applikation auch dann nicht zustande, wenn die Ratten 5 Tage hindurch mit steigenden

Tabelle 1

Gruppe	Zahl der Tiere	Askorbinsäuregehalt [mg/100 g Nebennierengewicht][1]	Wahrscheinlichkeit	Prozentuelle Abweichung von den Kontrollwerten %
Kontrollen (physiol. NaCl)	15	432 ± 22	–	100
Hexadimethrin-bromid	10	274 ± 24	$p < 0{,}001$	63
Substanz 48/80 (eine Injektion)	10	333 ± 14	$p < 0{,}01$	77
Substanz 48/80 (5tägige Vorbehandlung)	22	448 ± 14	$p > 0{,}05$	104
Substanz 48/80 (5tägige Vorbehandlung) + Hexadimethrinbromid	8	405 ± 32	$p > 0{,}05$	94
Phenergan + Hexadimethrin-bromid	8	227 ± 23	$p < 0{,}001$	53

[1] Mittelwert ± mittlerer Fehler

Dosen Substanz 48/80 vorbehandelt worden waren. In den mit Phenergan vorbehandelten Ratten bewirkte dagegen die i.v. Hexadimethrin-bromid-Zufuhr eine hochgradige Verringerung des Askorbinsäuregehaltes der Nebennieren.

Diskussion

Die Ergebnisse belegen, daß die intravenöse Gabe von Hexadimethrin-bromid außer einer Nebennierennekrose und Hemmung der Kortikoid-Produktion auch eine starke Verminderung des Nebennieren-Askorbinsäuregehaltes herbeiführt. Sie geht der Nekrose zeitlich voraus. Literaturangaben zufolge [11], [12] weist das darauf hin, daß Hexadimethrin-bromid die Funktion des Hypophysen-Nebennierenrinden–Systems zu aktivieren vermag, d. h. eine ACTH-Freisetzung auslöst.
Nach Kimura et al. [4], [5] kommt es durch Hexadimethrin-bromid zur Histaminmobilisation. Demzufolge könnte die durch Hexadimethrin-bromid bewirkte Askorbinsäureverminderung in den Nebennieren auch indirekt über die Histaminfreisetzung erfolgt sein. Das umso mehr, als das Histamin bzw. die Verbindung 48/80 bei Ratten eine gesteigerte ACTH-Sekretion und Verminderung des Askorbinsäuregehaltes der Nebennieren hervorrufen [3], [7], [8], [12], [14], [15]. Da bei den mittels Substanz 48/80 histamindepletierten Ratten die den Askorbinsäuregehalt der Nebennieren herabsetzende Wirkung des Hexadimethrin-bromids nicht eintrat, scheint sie indirekt unter Zwischenschaltung des Histamins zustande zu kommen. Dagegen spricht auch nicht, daß das Antihistaminikum Phenergan diese Askorbinsäuredepletion nicht zu verhindern vermochte. Es ist nämlich bekannt, daß Phenergan im akuten Versuch eine ACTH-Ausschüttung induziert und dadurch den Askorbinsäuregehalt der Nebennieren herabsetzt [9].

Literatur

[1] Carroll, R., K. Kovács u. E. Tapp: Lancet **2**, 921 (1964).
[2] Feldberg, W. u. J. Talesnik: J. Physiology **120**, 550 (1953).
[3] Halpern, B. N. u. S. A. Benos: Bull. Acad. suisse Sci. med. **8**, 110 (1952).
[4] Kimura, E. T., P. R. Young u. G. H. Barlow: Proc. Soc. exp. Biol. Med. **111**, 37 (1962).
[5] Kimura, E. T., P. R. Young u. R. K. Richards: Proc. Soc. exp. Biol. Med. **107**, 19 (1961).
[6] Kovács, K. u. Sz. Szeberényi: Naturwissenschaften **52**, 662 (1965).
[7] Nasmyth, P. A.: J. Physiology **112**, 215 (1951).
[8] Nasmyth, P. A.: Brit. J. Pharmacol. Chemotherapy **10**, 51 (1955).
[9] Rezabek, K. u. Z. Votava: Arch. int. Pharmacodynam. Thérap. **108**, 279 (1956).
[10] Roe, J. H. u. C. A. Kuether: J. biol. Chemistry **147**, 399 (1943).
[11] Sayers, G.: Physiol. Rev. **30**, 241 (1950).
[12] Sayers, G. u. M. Sayers: Endocrinology **40**, 265 (1947).
[13] Sayers, M. A., G. Sayers u. L. A. Woodbury: Endocrinology **42**, 379 (1948).
[14] Schönbaum, E. u. W. G. B. Casselman: Canad. J. Biochem. Physiol. **36**, 571 (1958).
[15] Schönbaum, E., W. G. B. Casselman u. R. E. Large: Canad. J. Biochem. Physiol **37**, 399 (1959).

Summary

K. Kovács and M. A. Dávid: Effect of hexadimethrine bromide on the ascorbic acid level in rat adrenals

Administration of hexadimethrine bromide (5 mg intravenously) or of compound 48/80 (0,1 mg intraperitoneally) reduced the ascorbic acid level in rat adrenals. The effect induced by hexadimethrine bromide could be prevented by pretreating the animals with substance 48/80, whereas phenergan left it unchanged. The results indicate that the reduction of the ascorbic acid level by hexadimethrine bromide goes in an indirect by way liberation of histamine.

Резюме

К. Ковач и М. А. Давид: Влияние гексадиметринбромида на содержание в надпочечниках крыс аскорбиновой кислоты

Введение гексадиметринбромида (5 мг внутривенно) или препарата 48/80 (0,1 мг внутрибрюшинно) снижает уровень аскорбиновой кислоты в надпочечниках крыс. Эффект, вызванный гексадиметринбромидом, может быть устранен предварительной обработкой животных препаратом 48/80, в то время как фенэрган не оказал никакого действия. Результаты свидетельствуют о том, что снижение уровня аскорбиновой кислоты гексадиметрином происходит прямым путем благодаря высвобождению гистамина.

Acta biol. med. german., Band 17, Seite 48–53 (1966)

Aus dem Physiologischen Institut der Friedrich-Schiller-Universität Jena
(Direktor: Prof. Dr. F. Schwarz)

Membranpotentialmessungen mit Mikroelektroden an Warmblütermuskeln unter den Bedingungen der Ischämie

M. Kunze

(Eingegangen am 16. 2. 1966)

Zusammenfassung

Anknüpfend an Arbeiten anderer Autoren über biochemische, histologische und funktionelle Veränderungen bei akuter Ischämie wurde an Ratten das Verhalten der Membranpotentiale bei mehrstündiger Ischämie der Skelettmuskulatur untersucht. Es fanden sich Potentialverringerungen auf 51, 37 und 30% der Ausgangswerte nach 3-, 4- und 5stündiger Ischämie. Der Potentialabfall war relativ gleichmäßig. Bei einer die Dauer von 4 Std. überschreitenden Ischämie konnte eine Restitution der Membranpotentiale während der Untersuchungszeit nicht gefunden werden.

Verschiedene Kliniker geben eine Wiederbelebungszeit der Skelettmuskulatur von 1½ bis maximal 3 Std. unter den Bedingungen der Ischämie (Esmarchsche Blutleere) an (Schmitt [1], Holle und Sonntag [2], Hellner und Mitarbeiter [3]). Eine diese Zeit überschreitende Anoxie führt zu irreversiblen strukturellen und funktionellen Veränderungen (ischämische Kontraktur), die besonders von Chirurgen gefürchtet werden. In den bisher vorliegenden experimentellen Arbeiten zu diesem Thema wurde vorwiegend über biochemische Untersuchungen berichtet, wobei die Veränderungen energiereicher Phosphorverbindungen und des Glykogens besondere Berücksichtigung fanden (Tremolieres und Mitarbeiter [4], Threlfall und Stoner [5], Maison [6]). Es ist bekannt, daß beim Erlöschen des aeroben Stoffwechsels die anaerobe Glykolyse als einzige Energiequelle in Frage kommt. Diese kann nicht über längere Zeit die ungestörte Zellfunktion aufrechterhalten, so daß in Abhängigkeit vom Energiebedarf, von der Sauerstoffreserve und der glykolytischen Aktivität des Organs die Zellfunktion zum Erliegen kommt. Da das Membranpotential vom Zellstoffwechsel abhängig ist, stellt es auch einen Indikator für seinen Funktionszustand dar. Deshalb war es interessant, die Veränderungen des Ruhepotentials in Abhängigkeit von der Zeit unter anoxämischen Bedingungen bei lebenden Tieren zu untersuchen.

Methodik

Die Untersuchungen wurden an weiblichen Albinoratten zwischen Mai und Dezember 1965 durchgeführt. Die Tiere wurden vor Beginn der Versuche mit Äthylurethan (120 mg/100 g Körpergewicht i.p.) narkotisiert. Es erfolgte eine vorsichtige Freilegung der linken Oberschenkelmuskulatur in einem Gebiet von etwa 2×1 cm. Nach Präparation der Faszie wurden 25 Membranpotentiale gemessen. Dann wurde nach vorherigem Anheben und Ausstreichen der Muskulatur mit Hilfe eines Gummischlauches am proximalen Ende des Oberschenkels abgebunden und in Abständen von 5 oder 15 min wurden je 5 Messungen vorgenommen. Die Abbindungen wurden nach 40 min, 3, 4 und 5 Std. gelöst und weitere Potentialmessungen bis 1 oder 1½ Std. danach vorgenommen. Zu einer Versuchsgruppe gehörten jeweils 10 Tiere.
Die Membranpotentiale wurden mit Glasmikroelektroden nach Ling und Gerard [7] abgeleitet, die nach der von Volkmer [8] beschriebenen Methode hergestellt wurden. Die Verstärkung und Messung der Potentiale erfolgte mit dem *p*H-Meßverstärker MV 11 (Clamann und Grahnert, Dresden), der über chlorierten Silberdraht mit den Ableitelektroden verbunden war. Die indifferente Elektrode bestand aus einer mit KCl-Agar gefüllten Kapillare mit weiter Öffnung. Sie wurde in Warmblüter–Ringer-Lösung zwischen Fell und Faszie nahe der Schnittfläche eingetaucht.
Während der Versuche erfolgte eine wiederholte Kontrolle der Elektrodenwiderstände in der bereits von Schwarz und Mitarbeitern [9] geschilderten Weise.
Die statistische Auswertung der Befunde wurde nach dem *t*-Test vorgenommen. Als statistisch signifikant gilt $P < 0{,}05$.

Ergebnisse

Wie aus Abbildung 1 ersichtlich ist, führt die akute Ischämie der Muskulatur zu einer Senkung des Membranpotentials. Diese ist bereits nach 15 min deutlich ($P < 0{,}01$). Nach 3-stündiger Ischämie betragen die Ruhepotentiale nur noch 51% der Ausgangswerte. Bei Wiederherstellung der Muskeldurchblutung sind erst nach etwa 1 Std. normale Membranpotentiale meßbar. Nach 4stündiger Ischämie betragen die durchschnittlichen Ruhepotentiale etwa 37% der Normalpotentiale. Die Erholungszeit, d. h. die Zeit von der Beendigung der Ischämie bis zum Auftreten normaler Potentiale, ist aber verlängert und beträgt 90 min. Nach 5stündiger Drosselung der Blutzufuhr kommt es zum Absinken der Potentiale auf 26 mV, das sind etwa 30% des Ausgangswertes. Jetzt ist der Wiederanstieg der Potentiale nach Lösen der Abbindung stark verzögert. Sie erreichen nach 1 Std. erst 62% der Normalwerte. Stichprobenweise durchgeführte Versuche ergaben, daß auch nach 2 und 3 Std. keine weitere Erhöhung der Ruhepotentiale eingetreten war.
Wie aus den Kurvenverläufen ersichtlich ist, fallen nach der Abbindung die Ruhepotentiale in den ersten 15 min wesentlich steiler ab als in den späteren Intervallen. Während in der ersten Viertelstunde in den 3 Versuchen der Potentialabfall 12,4, 14,2 und 17,4 mV betrug, wurden in den weiteren Intervallen keine Werte über 7,2 mV gemessen. Um diesen Potentialabfall in den ersten 15 min zeitlich genauer zu fassen, führten wir an 10 Ratten Messungen in Abständen von 5 min nach Abbindung durch (Abb. 2). Sie ergaben eindeutig, daß der vorher beobachtete starke Potentialabfall in den ersten 5 min eintritt und später ein gleichmäßiges Absinken zu verzeichnen ist.

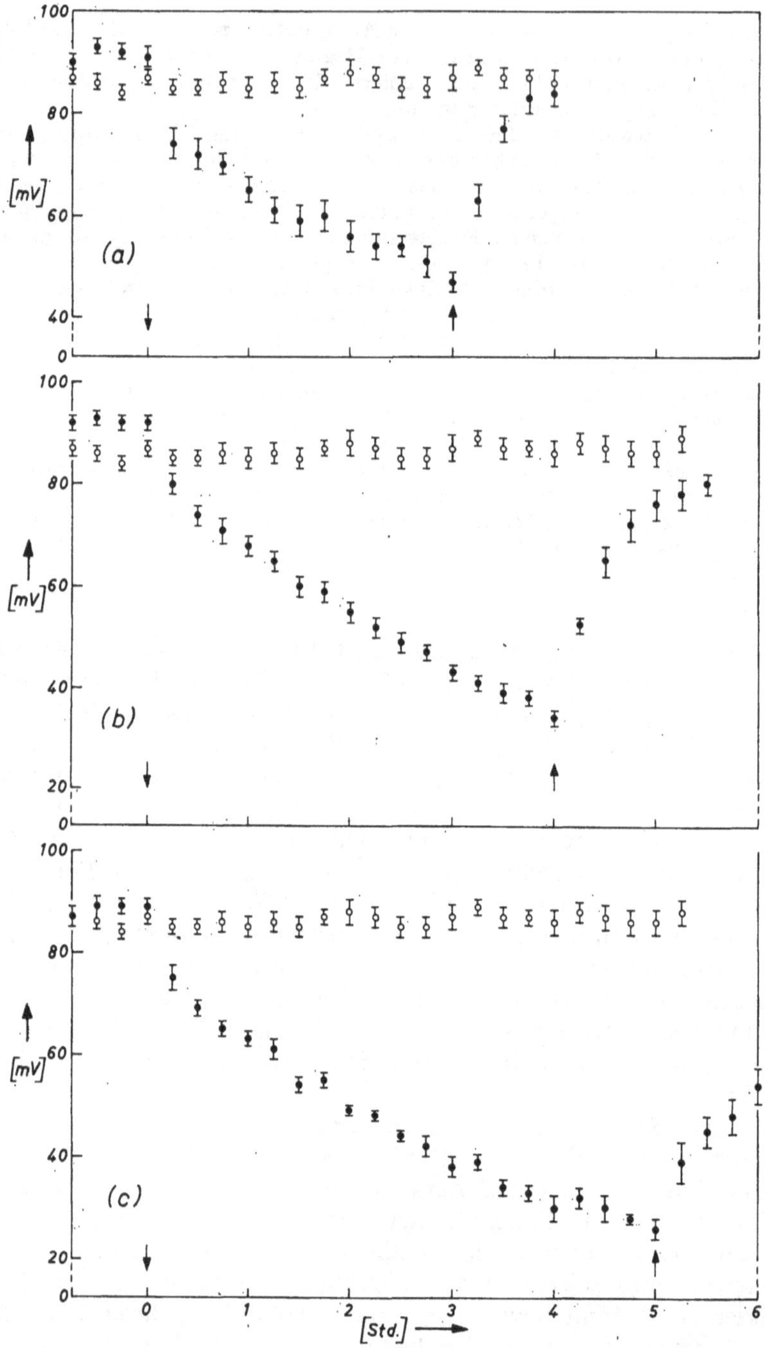

Abb. 1. Unterschrift siehe nebenstehend

Abb 2. Verhalten des Membranpotentials in Abhängigkeit von der Zeit bei einer Ischämie von 40 min Dauer, sonst wie in Abbildung 1

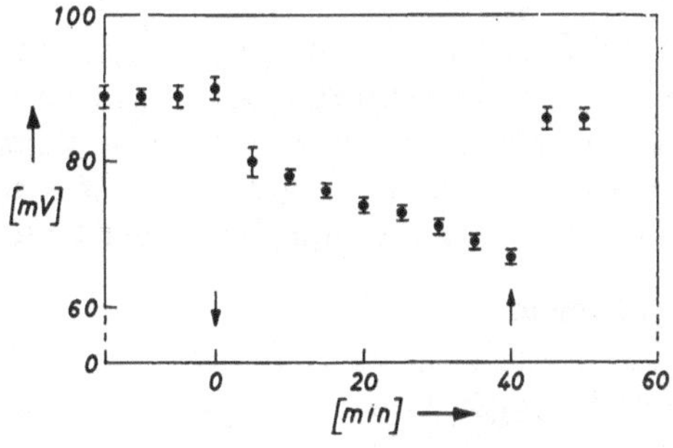

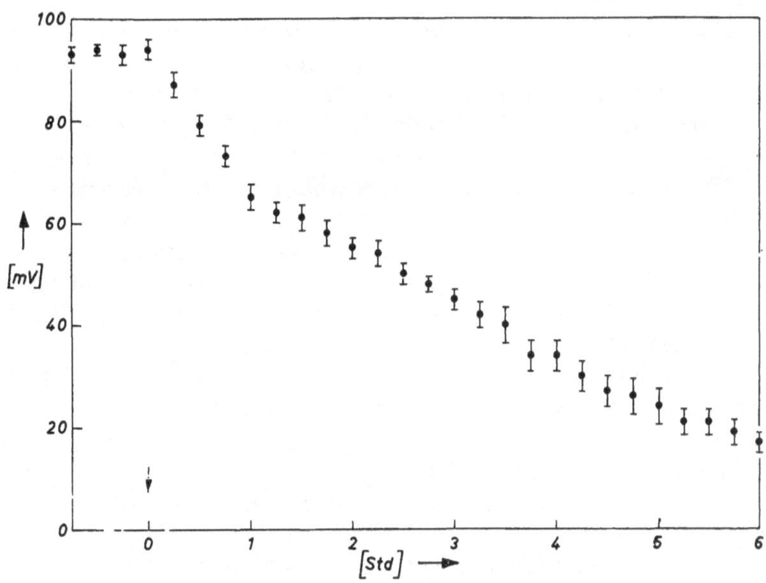

Abb. 3 Membranpotential bei Tötung der Tiere durch Ausbluten (↓), sonst wie in Abbildung 1

Abb 1 Verhalten des Membranpotentials in Abhängigkeit von der Zeit.

Weiße Punkte Potentiale der Kontrollgruppe, schwarze Punkte Potentiale bei (a) 3stündiger, (b) 4stündiger und (c) 5stündiger Ischämie, 45 min lang Messungen vor dem Abbinden Die Querstriche über und unter den Meßpunkten geben den mittleren Fehler des Mittelwertes an. ↓ = Abbinden der Gefäße, ↑ = Lösen der Abschnürung. Abszissen Zeit, Ordinaten Ruhepotential

Außerdem interessierte uns noch die Frage, ob die Veränderungen, die bei der Ischämie zu einer Reduzierung der Membranpotentiale führen, mit denen nach Tötung der Tiere übereinstimmen. Die Ergebnisse dieser Messungen bis 6 Std. nach Tötung der Tiere durch Ausbluten sind aus Abbildung 3 ersichtlich. Es fehlt der bei den Ischämieversuchen erwähnte steile Potentialabfall in den ersten 5 min. Ein Vergleich der stündlichen Potentialabfälle in allen Versuchen ergibt keine wesentlichen Abweichungen der Werte nach Tötung der Tiere von denen bei Ischämie.

Diskussion

Das Absinken von Membranpotentialen bei Warmblütern unter Anoxiebedingungen haben bereits Cresse und Mitarbeiter [10] nachgewiesen. Sie hatten an Rattenzwerchfellmuskeln eine Erniedrigung der Ruhepotentiale bis auf 15 mV gefunden. Da die Ergebnisse von isolierten, O_2- oder N_2-durchströmten Präparaten gewonnen wurden und lediglich 3 Std. nach Versuchsbeginn gemessen wurde, ist eine vollkommene Übereinstimmung nicht unbedingt zu erwarten. Gegenüber den Untersuchungen von Küchler und Mitarbeitern [11], die an Froschskelettmuskeln unter Anoxie einen Membranpotentialabfall von etwa 15% nach 6 Std. fanden, war der von uns bei Warmblütern erhaltene Potentialabfall (über 70 %) in derselben Zeit stärker ausgeprägt. Es wurde damit die Vermutung bestätigt, daß der Warmblütermuskel auch bezüglich des Ruhepotentials wesentlich empfindlicher auf O_2-Mangel reagiert.
Die von Klinikern empirisch angegebene Wiederbelebungszeit von 1½ bis höchstens 3 Std. bei kompletter Ischämie der Skelettmuskeln beruht überwiegend auf der Beurteilung der mechanischen Funktion. Die histologischen, biochemischen und funktionellen Veränderungen sind experimentell zum Teil von Kovach und Mitarbeitern [12] ohne Berücksichtigung des elektrischen Membranverhaltens bei Ratten untersucht worden. Sie fanden, daß Stoffe, die unter anaeroben Bedingungen an der Energiebereitstellung und -verteilung beteiligt sind (Glykogen, ATP), nach 3stündiger Ischämie deutlich in ihrer Konzentration verringert sind und zum Teil erst nach 1—2 Wochen wieder einen normalen Muskelspiegel aufweisen. Ebenso waren die mechanische Leistungsfähigkeit und die histologischen Veränderungen erst nach 2 Wochen wieder normalisiert. Vergleicht man diese Befunde und die biologischen Untersuchungsergebnisse anderer Autoren (Tremolieres und Mitarbeiter [4], Threlfall und Stoner [5], Thorn und Mitarbeiter [13]) mit den von uns gefundenen Membranpotentialwerten während der Ischämie, so ergibt sich eine Ähnlichkeit im Potentialabfall mit dem Verhalten des ATP-, Phosphorkreatin- und Glykogengehaltes. Auffällig sind nur die wesentlich länger feststellbaren Konzentrationsveränderungen der Phosphorverbindungen und des Glykogens nach 3stündiger Ischämie im Gegensatz zu der relativ schnellen Wiederherstellung des Membranpotentials. Es konnte auch für das Ruhepotential die Tatsache bestätigt werden, daß nach einer länger als 4 Std. dauernden Ischämie irreversible Veränderungen zu erwarten sind.

Eine größere Klarheit der Zusammenhänge von Prozessen, die an der Aufrechterhaltung einer gegen das Diffusionsgefälle gerichteten Ionenverteilung an der Membran und damit ihrer elektrischen Ladung beteiligt sind, müßte sich bei vergleichenden Untersuchungen von Elektrolyten, energiereichen Phosphaten und Membranpotential am selben Objekt unter den Bedingungen der Ischämie finden lassen. Diese Untersuchungen sollen in Zukunft durchgeführt werden.

Literatur

[1] Schmitt, W.: Allgemeine Chirurgie, Johann Ambrosius Barth-Verlag, Leipzig, 1955, S.122.

[2] Halle, F. u. E. Sonntag: Grundriß der gesamten Chirurgie, 7. Aufl. Springer-Verlag, Berlin-Göttingen-Heidelberg 1960, S. 123.

[3] Hellner, H., R. Nissen u. K. Vossschulte (Hrsg.): Lehrbuch der Chirurgie, 4. Aufl. Georg Thieme-Verlag, Stuttgart 1964, S. 29.

[4] Tremolieres, I., R. Deracke u. G. Griffaton: Bull. Soc. Chim. biol. **37**, 335 (1955).

[5] Threlfall, C. J. u. H. B. Stoner: Brit. J. exp. Pathol. **38**, 339 (1957).

[6] Maison, S. M.: Vopr. med. Chim. **2**, 133 (1956).

[7] Ling, G. u. R. W. Gerard: J. cellular comparat. Physiol. **34**, 383 (1949).

[8] Volkmer, H.: Acta biol. med. german. **15**, 348 (1965).

[9] Schwarz, F., M. Biedermann u. H. Volkmer: Acta biol. med. german. **15**, 608 (1965).

[10] Cresse, R., N. W. Schooles u. W. J. Whalen: J. Physiology **140**, 301 (1958).

[11] Küchler, G., J. Neudel u. L. Zett: Pflügers Arch. ges. Physiol. Menschen Tiere **278**, 395 (1963).

[12] Kovach, A. G. B., L. Takacs, M. T. Szabo, L. Takacs-Nagy, G. Zachariev u. I. Hamori: Acta physiol. Acad. Sci. hung. **10**, 313 (1956).

[13] Thorn, W., J. Heimann, B. Müldener u. R. Gercken: Pflügers Arch. ges. Physiol. Menschen Tiere **265**, 34 (1957).

Anschrift des Verfassers: Dr. Manfred Kunze, Physiologisches Institut der Universität Jena, 69 Jena, Teichgraben 8.

Summary

M. Kunze: Measurements of the membrane potential of ischemic rat muscles using micro-electrodes

Membrane potentials of rat skeletal muscles in ischemia of several hours have been studied basing on works of other investigators dealing with biochemical, histologic and functional changes in acute ischemia. Potentials decreased to 51, 37, and 30 percent of the initial levels after a 3, 4 and 5 hours' ischemia, the decrease being relatively smooth. In ischemia exceeding 4 hours no restitution of membrane potentials occurred throughout the period of investigation.

Резюме

М. Кунце: Измерения мембранных потенциалов крысиных мышц в условиях ишемии при помощи микроэлектродов

Основываясь на работах других авторов о биохимических, гистологических и функциональных изменениях в острой ишемии, авторы изучали мембранные потенциалы крысиных скелетных мышц в условиях длительной ишемии. Обнаруживалось уменьшение потенциалов до 51, 37 и 30 процентов исходных величин при ишемии длительностью 3, 4 и 5 часов соответственно. Потенциалы уменьшались равномерно. При ишемии выше 4 часов не наблюдалось восстановление мембранных потенциалов за период исследований.

Acta biol. med. german., Band 17, Seite 54—60 (1966)

Aus dem Physiologischen Institut der Friedrich-Schiller-Universität Jena
(Direktor: Prof. Dr. F. Schwarz)

Die Reizwirkung mittelfrequenter Wechselströme auf den Musculus orbicularis oris des Menschen

F. Schwarz

(Eingegangen am 16. 2. 1966)

Zusammenfassung

Der Musculus orbicularis oris des Menschen wurde mit Wechselströmen zwischen 1,5 und 12 kHz gereizt.
Die einzelnen Reizimpulse bringen unterschwellige Erregungen hervor, die sich summieren (Gildemeister-Effekt). Es zeigt sich, daß eine von der Reizfrequenz und der Reizspannung abhängige Zahl von Perioden abgelaufen sein muß, bis eine Muskelzuckung zustande kommt.
Bei der Durchströmung mit stärkerem Wechselstrom kommt es zu einem Tetanus, der nach einigen Sekunden wieder abklingt, obwohl der Strom weiterhin fließt (Mittelfrequenzhemmung). Die Kontraktionsdauer ist von der Reizfrequenz und der Reizspannung abhängig.
Ein Vergleich der Befunde mit denen an motorischen Nervenfasern des Frosches zeigt, daß für beide Versuchsobjekte offenbar die gleichen Gesetzmäßigkeiten gelten.

In den letzten Jahren sind einige Mitteilungen über die Wirkung mittelfrequenter Wechselspannungen auf motorische Nerven und Muskeln des Frosches und der Ratte erschienen (Schwarz [1]—[5], Schwarz und Ehrig [6], Ehrig [7], Schwarz und Rüster [8], Schwarz und Koch [9], Schwarz und Volkmer [10]).
Die Arbeiten beschäftigen sich mit der Summation unterschwelliger lokaler Erregungen bis zum Ausklinken eines Aktionspotentials und mit dem Abklingen der überschwelligen Erregung, der Mittelfrequenzhemmung. Um sich zu vergewissern, daß die Befunde an isolierten Nerven- und Muskelpräparaten auch auf in situ belassene Organe übertragbar sind, ist es nützlich, Versuche am Menschen anzustellen.

Methodik

Sinusförmige Wechselspannungen wurden einem Generator nach Zeluff [11] entnommen. Durch Druck auf eine Taste wurde ein Zeitschalter geöffnet, der einen Wellenzug von vorher eingestellter Dauer passieren ließ. Der Wellenzug wurde nach Verstärkung an einem Ausgangsübertrager mit der gewünschten Spannung abgegriffen und dem Muskel zugeleitet.
Mittels der differenten Elektrode, die 0,25 cm² groß war, aus Kupferblech bestand und mit Mull umwickelt war, wurde der Musculus orbicularis oris vom Lippenrot der Unter-

lippe aus gereizt. Es wurde darauf geachtet, daß die differente Elektrode während einer Meßreihe am gleichen Ort blieb, da die Reizschwelle vom Reizort abhängig ist. Die indifferente Elektrode, ebenfalls aus Kupferblech und mit Mull umwickelt, wurde auf der Stirn befestigt. Beide Elektroden waren mit Kochsalzlösung angefeuchtet. Als Reizerfolg wurde einmal die eben sichtbare Zuckung registriert, einmal die Zeit bis zum Ende der Kontraktion gemessen. Die Reizfrequenzen waren 1,5, 2, 3, 6 und 12 kHz. 4 Versuchspersonen standen zur Verfügung.

Ergebnisse

Während des Ablaufs eines Wellenzuges kommt es zur *Summation* unterschwelliger lokaler Erregungen. Bei Schwellenspannung wurde die Länge des Wellenzuges bestimmt, die zum Entstehen einer fortgeleiteten Erregung nötig war, es wurde also die ,,Hauptnutzzeit" (,,HNZ") in dem auf den Wechselstrom übertragenen Sinn bestimmt (SCHWARZ [2]). Die Abbildungen 1 und 2 zeigen das Ergebnis. Die ,,Hauptnutzzeiten" bei verschiedenen Reizfrequenzen lassen sich zu Kurven ordnen, die nach höheren Reizfrequenzen hin ansteigen und bei 6 kHz einen Knick aufweisen. Nach 12 kHz

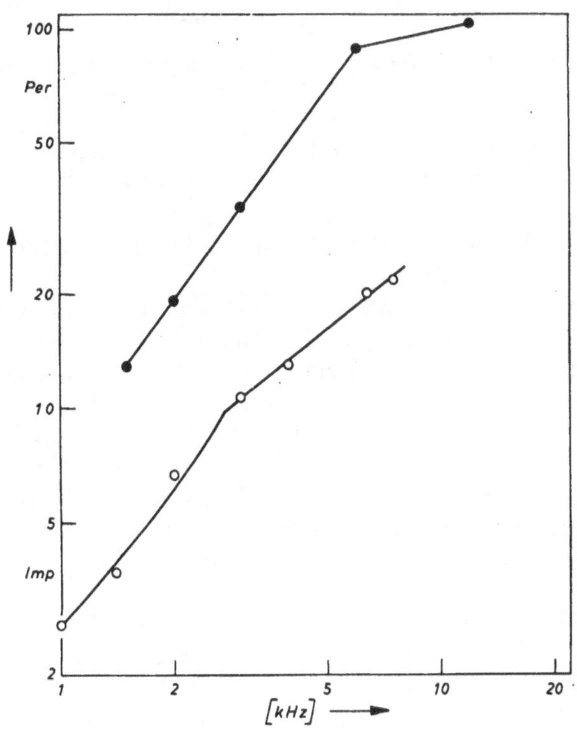

Abb. 1. ,,Hauptnutzzeiten", Befunde der Versuchspersonen B und V zusammengenommen. Beziehungen zwischen Periodenzahl (Impulszahl) und Reizfrequenz.

Abszissen: Reizfrequenz; Ordinaten: Zahl der Perioden (Impulse). Schwarze Punkte: Musculus orbicularis oris des Menschen; weiße Punkte: motorische Nervenfaser des Frosches

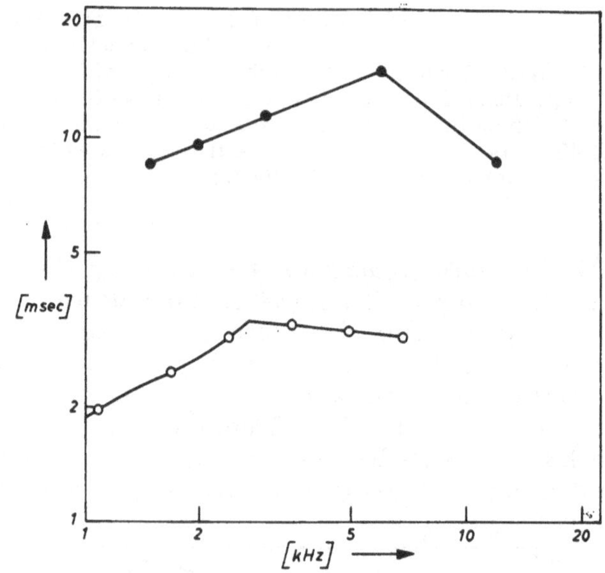

Abb. 2. „Hauptnutzzeiten", Befunde der Versuchspersonen B und V zusammengenommen. Beziehungen zwischen Dauer des Wellenzuges (der Impulsgruppe) und Reizfrequenz.

Abszissen: Reizfrequenz; Ordinaten: Dauer des Wellenzuges (der Impulsgruppe). Schwarze und weiße Punkte wie in Abbildung 1

hin verläuft die Kurve der Periodenzahl weniger steil nach oben, die der Dauer der Wellenzüge fällt ab.

Zum Vergleich sind die Befunde am RANVIERschen Schnürring der motorischen Nervenfaser des Frosches bei Reizung mit rechteckigen, nullsymmetrischen Wechselimpulsen (kathodisch-anodisch) in Abbildung 1 und 2 eingetragen worden (SCHWARZ und VOLKMER [10]). Auch hier nimmt die Zahl der Impulse pro „HNZ" mit steigender Frequenz zu, nach dem Knick bei 2,7 kHz ist der Kurvenverlauf weniger steil. Die Dauer der Impulsgruppen nimmt bis 2,7 kHz ebenfalls zu, um nach höheren Reizfrequenzen hin abzufallen. Die Zahl der Impulse pro „HNZ" und die Dauer der Impulsgruppen sind kleiner als die Periodenzahl und die Länge der Wellenzüge bei Reizung des menschlichen Muskels.

Die Streuungsbreite der einzelnen Meßergebnisse hat in den Versuchsreihen am menschlichen Muskel und an der motorischen Nervenfaser vom Frosch die gleiche Größenordnung. In Tabelle 1 sind die „Hauptnutzzeiten" der Wellenzüge und Impulsgruppen mit ihren Variabilitätskoeffizienten für 3 Versuchspersonen und zum Vergleich damit für Einzelfaserpräparate vom Frosch aufgeführt.

Von einem Wechselstrom bestimmter Frequenz ist eine um so kleinere Spannung zur Schwellenwirkung nötig, je länger der Wellenzug ist ([3]—[6], [8] bis [10]). Die Schwellenspannung eines Wellenzuges aus fünf Perioden

Tabelle 1

Statistischer Vergleich der „Hauptnutzzeiten“ bei Reizung des Musculus orbicularis oris des Menschen mit sinusförmigen Wechselspannungen und der motorischen Nervenfaser des Frosches mit nullsymmetrischen, rechteckigen Wechselimpulsen ($k-a$)

Frequenz kHz	Vp *B*		Vp *K*		Vp *V*		Froschnerv	
	$\bar{x}$	v_x	$\bar{x}$	v_x	$\bar{x}$	v_x	$\bar{x}$	v_x
1							2,12	34,4
1,5	8,2	30,0	9,3	16,5	9,1	15,3		
2	9,1	21,4	9,9	16,9	10,1	15,1	2,62	31,7
3	10,7	37,8	13,2	36,3	11,8	30,2	3,42	43,2
4							3,37	39,0
6	15,2	51,8	16,8	13,9	14,4	27,3		
7,5							2,92	47,7
12	8,6	27,5	8,1	33,3	8,7	11,9		

Anmerkung: Vp = Versuchsperson; $\bar{x}$ = „Hauptnutzzeit“ in msec, arithmetisches Mittel aus 12 (Mensch) und 22–27 (Frosch) Messungen, v_x = mittlere Abweichung der Einzelmessung in Prozent des Mittelwertes, Variabilitätskoeffizient

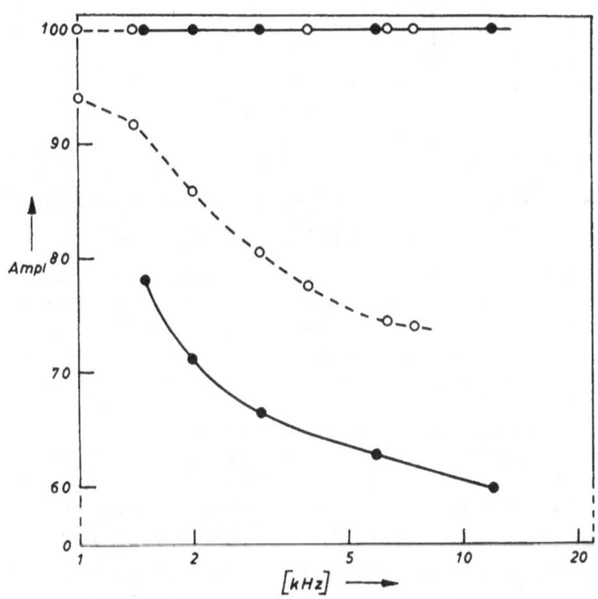

Abb. 3. Schwellenamplituden der Wellenzüge (Impulsgruppen) von der Länge der „Hauptnutzzeit“, Amplitude eines Wellenzuges von 5 Perioden oder eines einzigen Wechselimpulses = 100. Versuchspersonen B und V.

Abszissen: Reizfrequenz, Ordinaten: relative Amplituden. Schwarze und weiße Punkte wie in Abbildung 1

wurde = 100 gesetzt; mit kürzeren Wellenzügen wurde nicht gereizt, um die Einschaltwirkung zu umgehen (ROSENBLUETH [12], SCHWARZ [2], SCHWARZ und VOLKMER [10], WYSS [13]). Die Schwellenspannung eines Wellenzuges von der Länge der „Hauptnutzzeit" lag dann wesentlich unter 100 und war frequenzabhängig, wie Abbildung 3 zeigt.
Zum Vergleich ist wiederum das Ergebnis an der Einzelfaser aus dem Nervus ischiadicus des Frosches in Abbildung 3 eingetragen worden. Bei Reizung mit *k-a*-Wechselimpulsen liegen die Schwellenspannungen für „HNZ"-Impulsgruppen niedriger als bei Reizung mit e i n e m Wechselimpuls (= 100)[1].
Wie beim menschlichen Muskel sinkt die Schwellenspannung mit steigender Frequenz.
Die Summation unterschwelliger Erregungen führt bei größerer Reizstärke schneller (nach weniger Impulsen) zu einer Muskelzuckung. Die linke Seite von Abbildung 4 zeigt diese Intensität-Zeit-Beziehung bei 6 kHz.

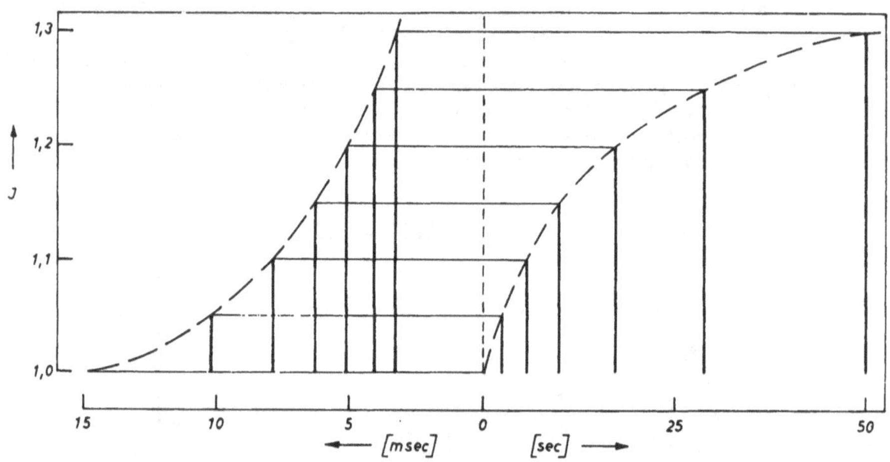

Abb. 4. Reizwirkung eines Wechselstroms von 6 kHz, Intensität–Zeit-Beziehungen. Links: „Nutzzeiten" des GILDEMEISTER-Effektes, *i–t*-Kurve (Summationskurve) gestrichelt. Versuchspersonen B und V. Rechts: Zusammenhang zwischen Reizstärke und Kontraktionszeit. Versuchsperson L. Abszissen: Zeit; Ordinaten: Reizstärke; Schwellenspannung (für „HNZ" und Zuckung) = 1

Die relative Schwellenspannung für einen Wellenzug von der Länge der „Hauptnutzzeit", die „Rheobase", ist 62,8 (Abb. 3); sie ist = 1 gesetzt worden. In der „Hauptnutzzeit" von 14,9 msec laufen 89 Perioden ab (Abb. 1 und 2). Bei größerer Amplitude genügen weniger Perioden, um eine Muskelzuckung auszulösen, die „Nutzzeit" ist kürzer; bei 1,3 kommt es bereits nach 20 Perioden zum Reizerfolg.

Die i-t-Kurve entspricht den Summationskurven, die sich bei der Reizung isolierter motorischer Nerven und Muskeln des Frosches und der Ratte ergeben haben ([3]—[6], [8]—[10]). Die rechte Seite der Abbildung 4 zeigt

[1] Da die Schwellenspannung für 1 Wechselimpuls = 5 Perioden = 100 gesetzt wurde, sind die Amplituden, die durch die beiden Kurven angegeben werden, nicht unmittelbar miteinander vergleichbar

den Zusammenhang zwischen der relativen Schwellenspannung und der Kontraktionsdauer bei 6 kHz. Es kommt bei der Reizung mit Wechselstrom zu einem Tetanus, der nach einigen Sekunden wieder abklingt, obwohl der Strom weiterhin fließt.

Je stärker der Wechselstrom ist, desto länger dauert der Tetanus an: bei der Schwellenspannung 1 (= „Rheobase") beobachtet man eine kurze Zuckung von etwa 0,1 sec Dauer, bei der Spannung 1,3 eine abklingende Kontraktion von 50 sec Dauer. LOTHHOLZ [14] hat diesen Effekt im Frequenzbereich von 4 bis 22 kHz untersucht.

Der gleiche Befund hatte sich auch bei Reizung des Nerv-Muskelpräparats (Nervus ischiadicus-Musculus gastrocnemius) vom Frosch (SCHWARZ [1]) und des gleichen Präparats von der Ratte (EHRIG [7]) ergeben: Die Kontraktionszeiten nahmen mit dem Ansteigen der Schwellenspannung zu.

Die 7 waagerechten Linien in Abbildung 4 verbinden die Effekte gleicher Amplituden: bei 1,0 eine „Hauptnutzzeit" von 14,9 msec mit einer Zuckung von etwa 0,1 sec Dauer, bei 1,05 eine „Nutzzeit" von 10,2 msec mit einer Kontraktion von 2,5 sec Dauer, usw.

Da offenbar für den Skelettmuskel des Menschen die gleichen Gesetzmäßigkeiten beim Zustandekommen einer Erregung gelten wie für die motorische Nervenfaser des Frosches, liegt es nahe, die Überlegungen und Schlußfolgerungen aus den Befunden am RANVIERschen Schnürring der Einzelfaser (SCHWARZ [15]) auch auf die Ergebnisse am menschlichen Muskel zu übertragen.

Literatur

[1] SCHWARZ, F.: Pflügers Arch. ges. Physiol. Menschen Tiere **250**, 343 (1948).
[2] SCHWARZ, F.: Pflügers Arch. ges. Physiol. Menschen Tiere **261**, 361 (1955).
[3] SCHWARZ, F.: Acta biol. med. german. **3**, 200 (1959).
[4] SCHWARZ, F.: Acta biol. med. german. **5**, 355 (1960).
[5] SCHWARZ, F.: Acta biol. med. german. **7**, 274 (1961).
[6] SCHWARZ, F. u. H. EHRIG: Pflügers Arch. ges. Physiol. Menschen Tiere **261**, 385 (1955).
[7] EHRIG, H.: Pflügers Arch. ges. Physiol. Menschen Tiere **262**, 127 (1956).
[8] SCHWARZ, F. u. M. RÜSTER: Acta biol. med. german. **2**, 321 (1959).
[9] SCHWARZ, F. u. W. KOCH: Acta biol. med. german. **5**, 561 (1960).
[10] SCHWARZ, F. u. H. VOLKMER: Acta biol. med. german. **15**, 283 (1965).
[11] ZELUFF, V.: Electronics (Amer.) December 1950, S. 120.
[12] ROSENBLUETH, A.: Ergebn. Physiol., biol. Chem. exp. Pharmakol. **47**, 24 (1952).
[13] WYSS, O. A. M.: Helv. physiol. pharmacol. Acta **23**, 107 (1965).
[14] LOTHHOLZ, H.: Acta biol. med. german. **12**, 167 (1964).
[15] SCHWARZ, F.: Acta biol. med. german. **16**, 625 (1966).

Anschrift des Verfassers: Prof. Dr. FRIEDRICH SCHWARZ, Physiologisches Institut der Universität Jena, 69 Jena, Teichgraben 8.

Summary

F. SCHWARZ: Stimulating effect of alternating currents of medium frequency on the Musculus orbicularis oris of man

Human Musculus orbicularis oris has been stimulated with alternating current in the frequency range from 1,5 to 12 kcps. The individual stimulating pulses produced

subthreshold excitations which are summing-up (GILDEMEISTER's effect). A certain number of periods depending on frequency and voltage of the stimulating current must have been passed, until a muscular contraction is stimulated.
Heavier alternating currents produced a tetanus which ceased after several seconds, though the current was still flowing (medium frequency inhibition). The time of contraction depends on frequency and voltage of the stimulating current. The results were compared with those obtained with motoric nerve fibres of the frog and both experimental objects found to be obviously subjected to common laws.

Резюме

Ф. Шварц: Раздражимость круговой мышцы рта человека при воздействии переменного тока средней частоты

Круговую мышцу рта человека подвергали раздражению переменным током частотой 1,5 до 12 гц. Отдельные раздражающие импульсы вызывают подпороговые возбуждения, которые суммируются (эффект Гильдемейстера). Мышца начинает сокращаться после протекания определенного числа периодов, зависимого от частоты и напряжения раздражающего тока.
Приложение более сильного переменного тока приводит к появлению столбняка, который исчезает спустя несколько секунд, несмотря на то, что подача тока продолжается (ингибирование средней частоты). Длительность сокращения зависима от частоты и напряжения раздражающего тока.
Сравнение этих данных с данными, полученными на моторных нервных волокнах лягушки показало, что оба опытных объекта очевидно подвергнуты одинаковым закономерностям.

Acta biol. med. german., Band 17, Seite 61–86 (1966)

Aus dem Physiologischen Institut der Friedrich-Schiller-Universität Jena
(Direktor: Prof. Dr. F. SCHWARZ)

Ionentransport in markhaltigen Nervenfasern

M. BRADL

(Eingegangen am 26. 2. 1966)

Zusammenfassung

Mit Hilfe von Leitfähigkeitsmessungen, chemischen Analysen und strukturellen Untersuchungen wird der Austausch von Chlor- und Kaliumionen zwischen Bündeln markhaltiger Nervenfasern und isotonischer Saccharoselösung mit und ohne Zusatz von 2,4-Dinitrophenol (DNP) studiert.
Für die zugehörigen Änderungen der intrazellulären Ionenkonzentration sowie für die ionalen Netto-Fluxe werden Differentialgleichungen abgeleitet, die den Zeitgang dieser Größen beschreiben.
Unvergiftete Nervenfasern geben Chlorionen in zwei Fraktionen, Kaliumionen dagegen in einer Fraktion ab. Bei DNP-Vergiftung der Präparate verschwindet die zweite Chlorfraktion, während sich in der ersten der Efflux in einen Influx umkehrt. Die Kaliumabgabe wird durch DNP nicht beeinflußt.
Es wird angenommen, daß der Chlorefflux auf einem aktiven Transportmechanismus mit energetischer Kopplung an den oxidativen Phosphatstoffwechsel beruht, während der Kaliumefflux einen passiven Diffusionsvorgang darstellt.

Wenn man Nervenfaserbündel in isotonischer Saccharoselösung badet, geben sie im Laufe der Zeit Ionen an diese Lösung ab. Wie in einer vorangegangenen Veröffentlichung gezeigt worden ist, kann man aus der damit verbundenen Änderung des elektrischen Leitwertes der Rohrzuckerlösung Rückschlüsse auf den zeitlichen Verlauf dieser Ionenabgabe ziehen (BRADL [4]).

Es ist möglich, die Leitfähigkeitsänderung auf mathematischem Wege in drei Komponenten mit unterschiedlichem Zeitverhalten zu zerlegen. Daraus läßt sich entnehmen, daß auch die Ionenabgabe aus den Nervenfasern in drei verschiedenen Fraktionen erfolgt. Unter Berücksichtigung der damit einhergehenden Änderung des Ruhepotentials der Präparate (BRADL [3]) konnten auf Grund der von HODGKIN [11], HODGKIN und HUXLEY [12], HODGKIN et al. [15] sowie HODGKIN und KATZ [16] entwickelten Ionentheorie der bioelektrischen Erscheinungen auch Aussagen über die Art der abgegebenen Ionen gemacht werden. Danach ist anzunehmen, daß in der ersten und dritten Fraktion Chlorionen, in der zweiten Fraktion dagegen Kaliumionen enthalten sind (BRADL [4]). Diese Schlußfolgerung hat jedoch nur qualitativen Charakter, denn schon aus Gründen der Elektroneutralität werden die einzelnen Fraktionen neben Chlor- und Kaliumionen wahrscheinlich auch noch andere Stoffe enthalten, die sich durch Ruhepotentialmessungen aber nicht identifizieren lassen. Es erhebt sich daher die Frage, ob es unter diesen Bedingungen möglich ist, die Gesetzmäßigkeiten der Chlor- und Kaliumabgabe isoliert zu erfassen. Die vorliegende Arbeit beschäftigt sich mit diesem Problem.

Desweiteren ist bekannt, daß Ionenverschiebungen zwischen dem intra- und extrazellulären Raum nicht immer nur passiv erfolgen, sondern vielfach auf einer Stoffwechselleistung der Zelle beruhen. In solchen Fällen kann man eine Beeinträchtigung der ionalen Transportvorgänge durch Stoffwechselgifte, z. B. durch 2,4-Dinitrophenol (DNP), erzielen. HODGKIN und KEYNES [17], [19] sowie CALDWELL et al. [7] haben z. B. unter Verwendung radioaktiver Isotope gezeigt, daß an markarmen Riesennervenfasern von Zephalopoden der Kalium- und Natriumaustausch zwischen dem Präparat und seiner Umgebung durch DNP empfindlich gestört wird. An markarmen und markreichen Nervenfasern von Warm- und Kaltblütern ist die Wirkung von DNP durch Messung des Membranpotentials untersucht worden (STRAUB [35], DETTBARN und STÄMPFLI [9], RITCHIE und STRAUB [27], SCHOEPFLE und BLOOM [31] sowie BÖHM und STRAUB [1]). Hierbei konnten aber nur Potentialänderungen von wenigen Millivolt registriert werden. Das trifft besonders für die markreichen Nervenfasern der Kaltblüter zu.

Es wurde nun die Ionenabgabe aus DNP-vergifteten Faserbündeln vom Nervus ischiadicus des Frosches durch Leitfähigkeitsmessungen am umgebenden Medium studiert. Auch über diese Untersuchungen soll im folgenden berichtet werden.

Methodik

1. Präparationstechnik

Zu den Versuchen dienten Faserbündel aus dem Nervus ischiadicus mittelgroßer Frösche (Rana temporaria und Rana esculenta). Bei der Präparation wurde die distale Aufspaltung des Nervenstammes in seine beiden Äste (Nervus tibialis und Nervus peronaeus) vollkommen freigelegt, der Nervus peronaeus möglichst weit distal durchtrennt und ein kleines Stück des Musculus gastrocnemius am Nervus tibialis belassen. Das Präparat kam auf eine mit isotonischer Saccharoselösung (213 mM) beträufelte Glasscheibe. Bei festgehaltenem Muskelstück wurde nun am Nervus peronaeus vorsichtig und gleichmäßig in proximaler Richtung gezogen. Die Bindegewebshülle riß dabei auf und der Nervenstamm teilte sich in zwei Äste. Unter einem binokularen Präpariermikroskop habe ich das zum Muskel führende Faserbündel bei Dunkelfeldbeleuchtung und 25facher Vergrößerung von Bindegewebsresten befreit, die sich meist sehr leicht am oberen oder am unteren Ende des Bündels mit gehalterten Nähnadeln und einer Präparierschere nach STÄMPFLI [33] mobilisieren ließen (vgl. BRADL [2]). Sie konnten dann mit einer Uhrmacherpinzette als einheitlicher Strang vom Präparat weggezogen und entfernt werden. Zum Schluß wurde noch der laterale, zum Peronaeus-Stumpf ziehende Nervenast an seinem proximalen Ende abgeschnitten. Bei einiger Übung lassen sich in der geschilderten Weise Faserbündel von mehreren Zentimetern Länge in 2—3 min freilegen.

2. Versuchsanordnung

Das Verfahren zur Leitfähigkeitsmessung ist schon in einer früheren Veröffentlichung ausführlich geschildert worden (BRADL [4]) und soll deshalb hier nur im Prinzip erläutert werden: Zur Aufnahme des Faserbündels diente ein Trog aus Vinidur, der mit 0,6 ml einer Testlösung gefüllt werden konnte. Das Präparat wurde quer über den Trog in dafür vorgesehene Schlitze gelegt. Die Abmessungen des Troges waren so beschaffen, daß die Flüssigkeit eine Faserstrecke von 5 mm Länge bedeckte. Um eine Verdunstung der Lösung zu verhindern, war die Anordnung in einer feuchten Kammer untergebracht. Die Ionenabgabe aus dem Faserbündel wurde durch Ermittlung des elektrischen Leitwertes der Testlösung erfaßt. Zu diesem Zweck war der Trog über Kohleelektroden an eine Widerstandsmeßbrücke angeschlossen, die mit Wechselspannung (Frequenz 1000 Hz) gespeist wurde. Die gemessenen Widerstandsdaten konnten dann in Leitwerte umgerechnet werden.

3. Testlösungen

Als Testflüssigkeit wurde Saccharoselösung von 213 mM benutzt, die bei einem Teil der Untersuchungen noch einen Zusatz von 0,2 mM DNP enthielt.

4. Versuche

a) Leitfähigkeitsmessungen

Zuerst wurde der Trog mit Testlösung beschickt und der Leitwert G_0 dieser Flüssigkeit bestimmt. Dann wurde ein Nervenfaserbündel in der geschilderten Weise präpariert, 10 min in 30 ml isotonischer Saccharoselösung gebadet und anschließend nochmals einige Sekunden durch Hin- und Herschwenken in frischer Rohrzuckerlösung ausgewaschen, um Ionen und andere Stoffe, die den Meßvorgang stören könnten, aus dem Extrazellulärraum der Fasern möglichst weitgehend zu entfernen. Zur Einschränkung des anhaftenden Flüssigkeitsfilms wurde das Präparat auf Filtrierpapier vorsichtig abgetupft und dann in die Meßanordnung eingelegt. Der Leitwert der Testlösung wurde nun alle 10 min registriert, bis eine Versuchsdauer von 200 min erreicht war. Da nur die Leitwertsänderung G^* interessierte, die durch den Ionenefflux aus dem Faserbündel entstand, wurde G_0 von den Leitwerten G_t, die nach Einlegen des Nerven in die Apparatur ermittelt worden sind, subtrahiert, denn es gilt die Beziehung

$$G_t = G_0 + G^*. \tag{1}$$

Die Experimente wurden abwechselnd mit DNP-Saccharoselösung und zum Vergleich mit Saccharoselösung ohne DNP-Zusatz durchgeführt. In jeder Serie wurden 10 Nerven untersucht.

b) Chemische Untersuchungen

Zur Untermauerung der Schlußfolgerungen, die aus den Leitwertsmessungen gezogen werden können, ist die Abgabe der Kalium-, Natrium- und Chlorionen aus den Nervenfaserbündeln noch durch einige orientierende chemische Analysen überprüft worden. Der Extrazellulärraum der Präparate wurde zunächst durch Vorbehandlung mit isotonischer Saccharoselösung von störenden ionalen Beimengungen befreit. Dann wurden die Bündel 200 min in den Testlösungen gebadet. Hierzu dienten Wannen aus Piacryl mit einem Fassungsvermögen von 5 ml. Die Abmessung der Gefäße war so beschaffen, daß die Badelösung hierbei eine Nervenstrecke von 10 mm Länge bedeckte. Nach Beendigung des Bades wurden jeweils 5 solcher Flüssigkeitsproben gesammelt und flammenphotometrisch auf ihren Kalium- und Natriumgehalt untersucht. Der Chlornachweis erfolgte durch Ausfällen mit Silbernitratlösung, nachdem die Proben mit verdünnter Salpetersäure versetzt worden waren.
Zum Vergleich wurden diese Untersuchungen auch mit den reinen Testlösungen durchgeführt.

c) Strukturelle Untersuchungen

Um den Ionenefflux aus den Nervenfasern und die damit verbundene Konzentrationsänderung der Ionen im Faserinneren erfassen zu können, war es notwendig, außer den Leitwertsänderungen des umgebenden Mediums auch den Durchmesser d und die Anzahl n der zu einem Bündel vereinigten Einzelfasern zu bestimmen.
Gelegentliche Beobachtungen hatten den Eindruck erweckt, daß die Nervenfaserbündel beim Aufbewahren in Rohrzuckerlösung etwas quellen. Es wäre denkbar, daß dabei der Durchmesser der Einzelfasern zunimmt. Um vergleichbare Verhältnisse zu schaffen, wurden deshalb die Präparate wie bei den Leitfähigkeitsmessungen zunächst 10 min in isotonischer Saccharoselösung gebadet und erst dann der Faserdurchmesser und die Faserzahl ermittelt.

d wurde an 125 Einzelfasern unter einem Mikroskop mit Hellfeldkondensor bei 400-facher Vergrößerung ausgemessen. Die Bestimmung des Durchmessers erfolgte im Internodium.

Zur Ermittlung der Faserzahl dienten zwei Methoden. Einmal wurde n aus d und dem Durchmesser D des gesamten Bündels näherungsweise errechnet. Obschon hierzu stark vereinfachende Annahmen über die Struktur des Nervenfaserbündels getroffen werden mußten, hat sich dieses Verfahren doch als sehr brauchbar erwiesen und soll deshalb anhand von Abbildung 1 erläutert werden. Dort ist der Querschnitt eines solchen Bündels in grob schematischer Form angedeutet. Danach kann man sich vorstellen, daß die einzelnen Nervenfasern (kleine Kreise) konzentrisch geschichtete „Schalen" (große gestrichelte Kreise) bilden. Da der Abstand zwischen den einzelnen „Schalen" gleich d ist, läßt sich die Anzahl m der „Schalen", die in einem Bündel vorkommen, aus d und D ermitteln:

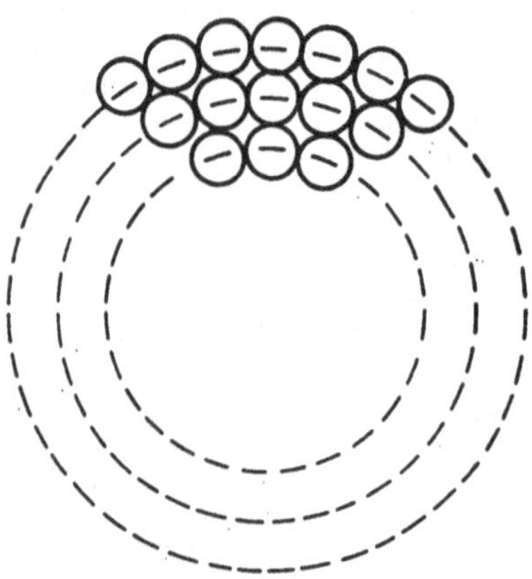

Abb. 1. Schema eines Nervenfaserbündels. Erklärung siehe Text

$$m = \frac{D}{2\,d} + 1\,. \qquad (2)$$

Die Zahl n_i der Fasern auf der i-ten „Schale" resultiert aus dem „Schalenumfang" x_i und dem Faserdurchmesser d zu

$$n_i = \frac{x_i}{d}\,. \qquad (3)$$

x_i kann man berechnen, wenn der „Schalendurchmesser" y_i bekannt ist:

$$x_i = y_i\,\pi\,. \qquad (4)$$

y_i ergibt sich aus der Formel

$$y_i = D - 2\,d\,(i - 1)\,, \qquad (5)$$

wobei die „Schalen" von außen nach innen fortschreitend gezählt werden und i alle positiven ganzen Zahlen von 1 bis m durchläuft. Aus der Formel (3) erhält man schließlich die Faserzahl n:

$$n = \sum_{i=1}^{i=m} n_i\,. \qquad (6)$$

Es kann vorkommen, daß sich die Gleichungen (2) und (3) nur angenähert auswerten lassen, weil die Brüche $D/2\,d$ und x_i/d nicht notwendigerweise ganze Zahlen darstellen, wie es für die Größen m und n_i zu fordern ist. In solchen Fällen muß entsprechend abgerundet werden.

Der für diese Rechnung benötigte Durchmesser des Faserbündels wurde auf mikroskopischem Wege bei 80facher Vergrößerung an 40 Präparaten bestimmt.

In einem zweiten Verfahren wurde n durch Auszählen von 10 Faserbündeln auch direkt ermittelt. Dazu diente ein Präpariermikroskop mit 25facher Vergrößerung und Dunkelfeldbeleuchtung.

Die Versuche wurden bei einer mittleren Raumtemperatur von 20 °C durchgeführt.

Ergebnisse

1. Leitfähigkeitsmessungen

Die Ergebnisse der Leitfähigkeitsmessungen veranschaulicht Abbildung 2. Die Leitwertsänderung G^* der jeweiligen Testlösung ist in Abhängigkeit von der Versuchsdauer t dargestellt. Es sind die arithmetischen Mittelwerte aus 10 Einzelmessungen eingetragen.

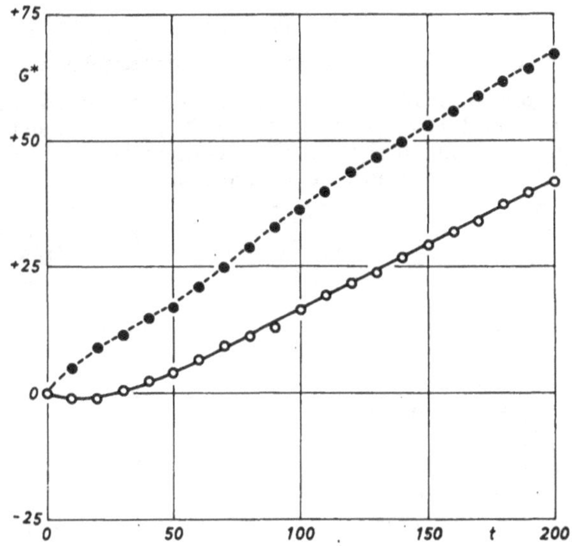

Abb. 2. Abhängigkeit der Leitwertsänderung G^* von der Versuchsdauer t. Punkte und gestrichelte Linie: Versuche mit Saccharoselösung; Kreise und durchgezogene Linie: Versuche mit DNP-Saccharoselösung. Ordinaten: G^* in $10^{-7}\ \Omega^{-1}$, Abszissen: t in min

Die unterschiedliche Wirkung der beiden Testlösungen ist sehr deutlich zu sehen. Zunächst fällt auf, daß die Leitwertsänderung unter DNP-Einfluß insgesamt geringer ist, als bei den Experimenten mit Saccharoselösung ohne Zusatz von DNP. Darüber hinaus weisen beide Kurven noch Unterschiede in der Gestalt auf. Die gestrichelte Linie verläuft ausschließlich im positiven Bereich der Ordinatenskala, sie besitzt bei $t \approx 50$ min eine Eindellung und ist im übrigen leicht nach oben durchgebogen. Im Gegensatz dazu zeigt die durchgezogene Kurve in ihrem Anfangsteil eine geringgradige Ausbuchtung nach unten; sie erreicht in diesem Gebiet sogar negative Ordinatenwerte und steigt ab $t \approx 60$ min linear an. Über die statistische Bearbeitung dieser Resultate orientiert Tabelle 1.

Es sollen noch die Leitwerte G_0 der reinen Testlösungen angegeben werden. G_0 beträgt bei isotonischer Saccharoselösung $(32{,}667 \pm 1{,}284)\ 10^{-7}\ \Omega^{-1}$, bei DNP-Saccharoselösung $(61{,}721 \pm 2{,}461)\ 10^{-7}\ \Omega^{-1}$. Es handelt sich dabei um die aus 10 Einzelwerten errechneten arithmetischen Mittelwerte und ihre mittleren Fehler.

2. *Mathematische Analyse der Leitwertsänderungen*

Für die Auswertung der Befunde ist es zweckmäßig, den Zusammenhang zwischen den Leitwertsänderungen G^* und der Versuchsdauer t in mathematischer Form darzustellen. Wie bereits gezeigt werden konnte (Bradl [4]), setzt sich G^* bei den Versuchen mit DNP-freier Rohrzuckerlösung additiv aus drei Komponenten G_1^*, G_2^* und G_3^* zusammen. Es gilt die Gleichung

$$G^* = G_1^* + G_2^* + G_3^*. \tag{7}$$

Tabelle 1

Statistische Angaben zu den Leitfähigkeitsmessungen.

Zeichenerklärung: S = Versuchsdauer in min; $\bar{x}$ = arithmetischer Mittelwert in $10^{-7}\ \Omega^{-1}$; σ_{x_i} = Standard-Abweichung in $10^{-7}\ \Omega^{-1}$; $\sigma_{\bar{x}}$ = mittlerer Fehler des arithmetischen Mittelwertes in $10^{-7}\ \Omega^{-1}$; t = STUDENT-Test; P = Überschreitungswahrscheinlichkeit in %

S	Versuche mit Saccharoselösung			$\leftarrow t$	$P \rightarrow$	Versuche mit DNP-Saccharoselösung		
	$\bar{x}$	σ_{x_i}	$\sigma_{\bar{x}}$			$\bar{x}$	σ_{x_i}	$\sigma_{\bar{x}}$
10	5,000	2,000	0,666			−0,889	3,775	1,253
20	8,812	1,782	0,577			−1,000	3,647	1,212
30	11,537	2,291	0,764			0,667	4,000	1,334
40	14,821	2,739	0,913			2,444	4,037	1,345
50	17,201	1,838	0,612	8,619	<0,1	4,222	4,301	1,435
60	21,222	2,598	0,866			6,556	3,782	1,257
70	24,812	2,958	0,986			9,333	4,062	1,353
80	29,000	3,507	1,145			11,333	4,472	1,490
90	32,889	3,606	1,200			13,222	3,873	1,292
100	36,600	4,837	1,612	10,326	<0,1	16,611	3,950	1,319
110	40,111	4,159	1,386			18,667	4,528	1,510
120	44,100	3,225	1,072			21,778	3,873	1,292
130	47,000	5,206	1,735			23,889	4,461	1,487
140	49,889	3,647	1,212			27,000	4,743	1,581
150	53,100	4,980	1,658	6,652	<0,1	29,333	5,814	1,936
160	56,000	5,225	1,776			32,111	6,557	2,182
170	58,826	5,595	1,863			34,250	5,916	2,093
180	62,000	5,701	1,900			37,625	6,760	2,390
190	64,235	6,356	2,119			39,875	8,270	2,924
200	67,289	5,958	1,985	6,886	<0,1	42,235	7,280	2,751

Die Abhängigkeit der einzelnen Summanden von der Zeit läßt sich durch folgende Ausdrücke beschreiben:

$$G_1^* = -\frac{\varphi}{a_g} e^{-a_g t} + \chi, \tag{8}$$

$$G_2^* = \psi t \tag{9}$$

und

$$G_3^* = \int \omega\, \tau^{c_g} e^{-b_g \tau}\, d\tau. \tag{10}$$

Den Ansatz (8) kann man noch vereinfachen, denn es besteht die Relation

$$\frac{\varphi}{a_g} = \chi\,. \tag{11}$$

Damit geht die Formel (8) in den Ausdruck

$$G_1^* = \chi\,(1 - e^{-a_g t}) \tag{12}$$

über.

Die Zeitkoordinaten wurden bei G_1^* und G_2^* mit t, bei G_3^* jedoch mit τ bezeichnet, weil G_3^* 40 min später als die beiden ersten Komponenten beginnt.

Die neue Variable τ wurde gemäß

$$\tau = t - 40 \tag{13}$$

eingeführt und G_3^* nicht als Funktion von t, sondern als Funktion von τ behandelt. Durch diese Maßnahme konnte der Beginn von G_3^* in den Nullpunkt der Zeitachse verlegt und somit die Rechnung vereinfacht werden (vgl. Bradl [3]). Die Größen φ, $\varkappa$, ψ, ω, a_g, b_g und c_g, die in den Ausdrücken (8), (9), (10), (11) und (12) vorkommen, stellen Konstanten dar.

Um die entsprechenden Verhältnisse bei Vergiftung der Nervenfaserbündel mit DNP zu klären, ist untersucht worden, aus welchen Funktionen sich die durchgezogene Kurve der Abbildung 2 zusammensetzt. Die dazu benutzten Rechenverfahren sind bei Bronstein und Semendjajew [5] angegeben. G^* besteht in diesem Falle nur aus zwei Komponenten G_1^* und G_2^*, von denen die erste dem Exponentialgesetz (12) folgt, während die zweite einen linearen Verlauf besitzt und durch die Funktion (9) wiedergegeben werden kann. Diese Komponenten ergänzen sich aber nicht wie beim unvergifteten Nerven additiv, sondern subtraktiv zu G^*. Mit der Gleichung

$$G^* = -G_1^* + G_2^* \tag{14}$$

erhält man eine Kurve, die der durchgezogenen Linie in Abbildung 2 sehr genau entspricht.

Tabelle 2 orientiert über alle Konstanten, die in den Ausdrücken (9), (10) und (12) vorkommen. Wie man sieht, hängen die Zahlenwerte dieser Größen davon ab, ob die Versuche mit DNP-freier oder mit DNP-haltiger Testlösung durchgeführt worden sind. Das trifft besonders für die Parameter der Komponente G_1^* zu, während sich die Konstante ψ unter dem Einfluß von DNP nur sehr wenig ändert.

Tabelle 2

Zusammenstellung der Konstanten, die sich bei der mathematischen Analyse der Leitfähigkeitsänderungen G^* ergeben haben

Konstante	Betrag der Konstanten		Dimension der Konstanten
	Versuche mit Saccharoselösung	Versuche mit DNP-Saccharoselösung	
χ	3,657	8,765	$10^{-7}\ \Omega^{-1}$
ψ	0,275	0,255	$10^{-7}\ \Omega^{-1}\ \mathrm{min}^{-1}$
ω	0,000246	—	$10^{-7}\ \Omega^{-1}\ \mathrm{min}^{-2}$
a_g	0,111	0,051	min^{-1}
b_g	0,063	—	min^{-1}
c_g	2,413	—	—

In Abbildung 3 ist der zeitliche Verlauf der Komponenten G_1^*, G_2^* und G_3^* dargestellt. Die ersten Komponenten (Abb. 3a) streben gemäß der Funktion (12) exponentiell einem konstanten Endwert zu. Dieser Endwert ist bei Behandlung der Präparate mit DNP ungefähr doppelt so groß, wie bei den

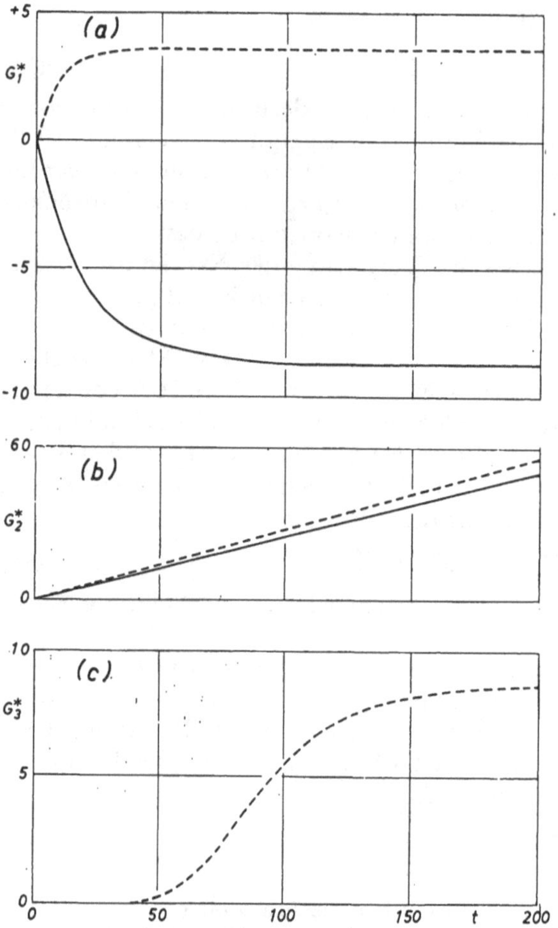

Abb. 3. Abhängigkeit der Leitwertskomponenten $G_1^*(a)$, $G_2^*(b)$ und $G_3^*(c)$ von der Versuchsdauer t. Gestrichelte Linien: Versuche mit Saccharoselösung; durchgezogene Linien: Versuche mit DNP-Saccharoselösung. Ordinaten: G^* in $10^{-7}\ \Omega^{-1}$, Abszissen: t in min

Experimenten mit unvergifteten Nervenfaserbündeln und wird von der durchgezogenen Linie auch wesentlich später erreicht als von der gestrichelten Kurve. Die Zeitdifferenz beträgt etwa 50 min. Wegen des unterschiedlichen Vorzeichens von G_1^* in den Gleichungen (7) und (14) liegt der gestrichelte Kurvenzug im positiven, die durchgezogene Linie im negativen Ordinatenbereich. Diese Darstellung zeigt sehr instruktiv, daß G_1^* im ersten Fall einer Leitfähigkeitszunahme, im zweiten Fall einer Leitwertsabnahme der Testlösung entspricht. Wie aus Abbildung 3b hervorgeht, hat DNP auf die Komponente G_2^* kaum einen Einfluß. Die ansteigenden Geraden liegen sehr dicht beieinander. Die dritte Komponente (Abb. 3c), die sich nur beim Experimentieren mit unvergifteten Faserbündeln nachweisen läßt, beginnt erst, nachdem die Nervenfasern bereits 40 min mit isotonischer Saccharoselösung behandelt worden sind. Sie strebt in S-förmigem Verlauf einem Sättigungswert zu, der bei $t \approx 200$ min erreicht wird.

3. *Chemische Untersuchungen*

Die Ergebnisse der orientierenden chemischen Untersuchungen sind in Tabelle 3 eingetragen. Wie man sieht, geben die Nervenfaserbündel beim Baden in den Testlösungen Kalium-, Natrium- und Chlorionen an das umgebende Medium ab. Der Kaliumaustritt ist ungefähr dreimal so groß wie die Natriumabgabe. Während der Efflux der Kationen von DNP kaum beeinflußt wird, reduziert dieses Stoffwechselgift den Efflux der Chlorionen

Tabelle 3

Ergebnisse der chemischen Untersuchungen.

Zeichenerklärung: A = Versuche mit Saccharoselösung; B = Versuche mit DNP-Saccharoselösung.

Die eckigen Klammern bedeuten die Konzentrationen der betreffenden Ionen in mMol/1000 ml. Bei den Fällungsreaktionen ist ein negatives Ergebnis durch —, ein positives Ergebnis durch + angedeutet. Die Intensität der Ausfällung wird durch die Anzahl der + veranschaulicht

	[K]		[Na]		Cl	
	A	*B*	*A*	*B*	*A*	*B*
I Testlösung vor Einlegen des Präparates	0,118	0,118	0,043	0,043	—	—
II Testlösung 200 min nach Einlegen des Präparates	0,146	0,148	0,053	0,054	++	+
Differenz zwischen I und II	0,028	0,030	0,010	0,011		

sehr deutlich. Schon die qualitative Fällungsreaktion läßt diesen Sachverhalt gut erkennen.

4. *Strukturelle Untersuchungen*

Die mikroskopischen Untersuchungen haben für die einzelne Nervenfaser einen Durchmesser von 14,963 ± 0,238 μ, für das Faserbündel einen solchen von 403,000 ± 7,000 μ erbracht. Es sind die arithmetischen Mittelwerte und ihre mittleren Fehler angegeben. Sie wurden im ersten Fall aus 125, im zweiten Fall aus 40 Einzelbeobachtungen gewonnen. Aus diesen Resultaten folgt unter Zugrundelegung der Formeln (2) bis (6), daß ein Bündel im Durchschnitt 570 Einzelfasern enthält. Durch Auszählen von 10 Bündeln konnte eine mittlere Faserzahl von 585 ± 29 festgestellt werden. Die Abweichung dieser beiden Mittelwerte voneinander ist zufallsbedingt, denn ihre Differenz beträgt nur 15 und liegt somit weit innerhalb der Fehlerbreite ± 29.

Diskussion

1. *Veränderung der intrazellulären Ionenkonzentration*

Die Leitfähigkeitsänderungen lassen sich deuten, wenn man annimmt, daß die Ionenkonzentration der Testlösungen während der Experimente größer oder kleiner wird. Eine Verdunstung von Flüssigkeit kommt als Ursache hierfür nicht in Betracht, weil das Meßgefäß in einer feuchten Kammer untergebracht worden war. Die Konzentrationsänderungen können also nur darauf beruhen, daß die Nervenfaserbündel entweder Ionen an die Badelösung abgeben oder vom Außenmilieu aufnehmen. Im ersten Fall müßte eine Leitwertszunahme, im zweiten Fall eine Leitfähigkeitsabnahme der Testlösung resultieren.

Die Leitwertsvergrößerung, die sich ergibt, wenn als Testflüssigkeit isotonische Saccharoselösung ohne DNP-Zusatz benutzt wird, ist bereits in diesem Sinne gedeutet worden (BRADL [4]). Es wurde angenommen, daß die Ionen entsprechend den drei Komponenten von G^* die Nervenfaserbündel in drei verschiedenen Fraktionen verlassen, wobei die erste und dritte Fraktion Chlorionen, die zweite Fraktion dagegen Kaliumionen enthält.

Bei Vergiftung des Nerven mit DNP läßt sich G^* nur in zwei Komponenten zerlegen. Daraus kann man schließen, daß auch der Ionenaustausch zwischen dem Präparat und seiner Umgebung hier nur in zwei Fraktionen erfolgt.

Da G_1^* unter DNP-Einfluß eine Leitfähigkeitsabnahme der Testlösung anzeigt, nimmt das Faserbündel in diesem Fall Ionen des Außenmediums auf. Wie Ruhepotentialmessungen[1] an den Präparaten ergeben haben, geht der Leitwertsverminderung eine ausgeprägte Membrandepolarisation parallel. Die Abnahme des Ruhepotentials und die Ionenaufnahme bei DNP-vergifteten Nervenfasern dürften deshalb eng miteinander zusammenhängen. Nach der von HODGKIN [11], HODGKIN und HUXLEY [12], HODGKIN et al. [15] sowie HODGKIN und KATZ [16] entwickelten Ionentheorie der bioelektrischen Erscheinungen wirkt eine Erhöhung der Chlorionenkonzentration im Zellinneren depolarisierend. Man kann daraus folgern, daß bei einem Nerven, der mit DNP vergiftet worden ist, die erste Fraktion wie beim unvergifteten Präparat Chlorionen enthält, die aber nicht aus dem Faserbündel ausströmen, sondern umgekehrt vom umgebenden Medium in das Bündel eindringen.

Nun konnte zwar in den Testlösungen vor Einlegen des Nerven kein Chlorid festgestellt werden (vgl. Tab. 3); es ist aber zu beachten. daß die zum Chlornachweis benutzte chemische Reaktion nach KARAOGLANOW [20] nur bis zu einer unteren Konzentration von ungefähr 2,707 μMol Chlorionen/1000 ml Lösung positiv ausfällt. Möglicherweise sind also in den Testflüssigkeiten Chlorionen in einer Menge enthalten, die unterhalb der angegebenen Nachweisgrenze liegt. Wegen des sehr kleinen Volumens der Nervenfasern könnte diese Menge ausreichen, um beim Vordringen in das Zellinnere dort eine merkliche Konzentrationsänderung zu bewirken. Das negative Ergebnis der chemischen Analyse ist deshalb kein stichhaltiges Argument gegen die Annahme eines Chlortransportes vom Außenmilieu in das Faserbündel hinein.

Wie Abbildung 3b zeigt, wird G_2^* und damit auch die zweite Ionenfraktion von DNP kaum beeinflußt. Desgleichen hat nach Tabelle 3 DNP auf die Abgabe von Kaliumionen aus dem Präparat praktisch keine Wirkung. Auch HODGKIN und KEYNES [17] konnten einen deutlichen DNP-Effekt auf den Kaliumefflux an markarmen Riesennervenfasern von Zephalopoden nicht feststellen. Diese Ergebnisse sprechen für die Annahme, daß in der zweiten Fraktion Kaliumionen enthalten sind.

Wie bereits erwähnt worden ist (BRADL [4]), könnten in dieser Fraktion auch noch Natriumionen vorkommen. Die Ergebnisse der flammenphotometrischen Untersuchungen scheinen diese Hypothese zu stützen. Dabei tritt aber das Natrium gegenüber dem Kalium mengenmäßig in den Hintergrund.

[1] Darüber soll in einer anderen Arbeit berichtet werden

Mit der dritten Ionenfraktion, die der Leitwertskomponente G_3^* entspricht, strömen vermutlich Chlorionen aus dem Nervenfaserbündel aus. Da G_3^* durch DNP unterdrückt werden kann, müßte demnach auch die Chlorabgabe der Präparate durch DNP zumindest eingeschränkt werden. Aus Tabelle 3 kann man entnehmen, daß diese Schlußfolgerung richtig ist.

Durch die vorliegenden Befunde werden die einleitend skizzierten Annahmen über die Abgabe von Chlor- und Kaliumionen aus Nervenfaserbündeln gestützt. Zusätzlich hat sich herausgestellt, daß DNP den Chloraustausch zwischen dem Präparat und seiner Umgebung stark beeinflußt. Um diesen qualitativen Resultaten eine quantitative Fassung zu geben, soll zunächst untersucht werden, welchen Gesetzen die mit dem Ionentransport verknüpften Änderungen der intrazellulären Chlor- und Kaliumkonzentration gehorchen. Diese Gesetze lassen sich aus dem zeitlichen Verhalten der Leitwertskomponenten G_1^*, G_2^* und G_3^* ableiten.

Wenn das Nervenfaserbündel beim Baden in der Testlösung die Ionensorte x abgibt, so wird die Konzentration dieser Ionen im Außenmilieu um C_x^* erhöht und der elektrische Leitwert der Flüssigkeit nimmt um G_x^* zu. Zwischen G_x^* und C_x^* besteht die Beziehung (BRADL [4])

$$G_x^* = \frac{1}{1000} \frac{q}{l} z_x l_x C_x^* , \tag{15}$$

wobei G_x^* in Ω^{-1} und C_x^* in Mol/1000 ml Lösung einzusetzen sind. q stellt den Querschnitt, l die Länge der Flüssigkeitssäule im Meßgefäß dar, z_x ist die Ladungszahl und l_x die Beweglichkeit der betreffenden Ionen. Durch Umwandeln der Gleichung (15) erhält man den Ausdruck

$$C_x^* = \frac{1000\, l}{q\, z_x\, l_x} G_x^* . \tag{16}$$

Die absolute Menge $|C_x^*|$ der vom Faserbündel abgegebenen Ionen in Mol resultiert, wenn der Ansatz (16) mit dem Faktor $ql/1000$ multipliziert wird:

$$|C_x^*| = \frac{l^2}{z_x\, l_x} G_x^* . \tag{17}$$

Für die weiteren Betrachtungen muß das Volumen V_N des Präparates bekannt sein, aus dem die Ionenmenge $|C_x^*|$ stammt. Das Gesamtvolumen des Bündels setzt sich praktisch nur aus den Volumina der Einzelfasern und der interfibrillären Spalträume zusammen. Die zwischen den Nervenfasern verlaufenden Bindegewebsfibrillen sind so dünn, daß sie unberücksichtigt bleiben können. Nach Befunden von SCHMIDT [30] erfolgt der Stoffaustausch zwischen den Spalträumen und dem Außenmilieu des Nerven viel zu rasch, um für die relativ langsamen Änderungen von G^* eine Rolle zu spielen. Außerdem ist der Interfibrillärraum durch Behandlung der Präparate mit isotonischer Saccharoselösung vor Versuchsbeginn hinreichend ausgespült worden. $|C_x^*|$ kann also nur von den Nervenfasern selbst abgegeben worden sein. Wenn man die Zahl n der im Bündel enthaltenen Fasern, den Faserdurchmesser d und die Länge λ der von der Testlösung bedeckten Nervenstrecke kennt, kann man V_N ausrechnen. Es gilt die Beziehung

$$V_N = n \frac{d^2}{4} \pi \lambda. \tag{18}$$

Die Konzentrationsänderung C^*_{ix} in Mol/1000 cm³ Nervenfaser, die der Efflux der Ionensorte x im Faserinneren erzeugt, liefert der Ansatz (17) bei Multiplikation mit dem Faktor $1000/V_N$:

$$C^*_{ix} = \frac{l^2}{z_x l_x} \frac{1000}{n \frac{d^2}{4} \pi \lambda} G^*_x . \tag{19}$$

Diese Formel gilt natürlich auch dann, wenn C^*_{ix} durch einen Ioneninflux verursacht wird.
Nach Gleichung (19) lassen sich aus den Funktionen G^*_1, G^*_2 und G^*_3 die Veränderungen C^*_{i1}, C^*_{i2} und C^*_{i3} der intrazellulären Ionenkonzentrationen bestimmen, sofern jeder Leitwertskomponente die Abgabe oder Aufnahme von nur einer Ionensorte durch den Nerven entspricht. Deshalb soll vorerst angenommen werden, daß in der ersten und dritten Ionenfraktion lediglich Chlorionen, in der zweiten Fraktion nur Kaliumionen vorkommen. Die Rechnung wird zeigen, ob diese Annahmen zutreffen.
Weil die Ladungszahlen der Chlor- und Kaliumionen beide gleich 1 sind, kann man aus dem Ansatz (19) die Faktoren

$$k_{Cl} = \frac{l^2}{l_{Cl}} \frac{1000}{n \frac{d^2}{4} \pi \lambda} . \tag{20a}$$

und

$$k_K = \frac{l^2}{l_K} \frac{1000}{n \frac{d^2}{4} \pi \lambda} . \tag{20b}$$

bilden. Mit den Formeln (20a) und (20b) ist es unter Berücksichtigung des Zusammenhanges (19) möglich, die Größen C^*_{i1}, C^*_{i2} und C^*_{i3} aus den Funktionen (12), (9) und (10) abzuleiten. Für die Versuche mit DNP-freier Saccharoselösung erhält man die Ausdrücke

$$C^*_{i1} = k_{Cl} \chi (1 - e^{-a_g t}) , \tag{21a}$$

$$C^*_{i2} = k_K \psi t \tag{21b}$$

und

$$C^*_{i3} = k_{Cl} \int \omega \tau^{c_g} e^{-t_g \tau} d\tau, \tag{21c}$$

bei Verwendung DNP-haltiger Rohrzuckerlösung gelten nur die Ansätze (21a) und (21b).
Für die numerische Auswertung der Gleichungen (21a), (21b) und (21c) müssen noch die in den Formeln (20a) und (20b) vorkommenden Konstanten bekannt sein. Die Beweglichkeiten der Chlor- und Kaliumionen sind bei

D'ANS und LAX [8] (bezogen auf unendlich verdünnte Lösungen und eine Temperatur von 18 °C) mit $l_{Cl} = 65{,}5\ \mathrm{cm}^2\ \Omega^{-1}$ und $l_K = 64{,}6\ \mathrm{cm}^2\ \Omega^{-1}$ verzeichnet. Für n wurde der errechnete Wert von 570 benutzt, weil dieses Ergebnis mit den Befunden anderer Autoren (RUD [28], SCHMIDT [29]) besser übereinstimmt, als das durch Auszählen der Faserbündel erhaltene Resultat. Des weiteren betragen $l = 2$ cm, $\lambda = 0{,}5$ cm und $d = 14{,}963 \cdot 10^{-4}$ cm. Unter diesen Voraussetzungen sind die Konzentrationsänderungen C_{i1}^*, C_{i2}^* und C_{i3}^* berechnet und in Abbildung 4 als Funktionen der Versuchsdauer t dargestellt worden. Um die „biphasische" Abgabe der Chlorionen aus dem normalen Nerven noch besonders zu verdeutlichen, ist in Abbildung 5 die Summe aus C_{i1}^* und C_{i3}^* in Abhängigkeit von t wiedergegeben. Nach der Formel (19) sind die intrazellulären Konzentrationsänderungen den entsprechenden Leitwertskomponenten der Testlösung direkt proportional. Dieser Sachverhalt wird auch deutlich, wenn man die Funktionen (21 a), (21 b) und (21 c) jeweils mit den Ausdrücken (12), (9) und (10) vergleicht. Die Kurven der Abbildung 4 besitzen daher denselben Zeitgang wie die zugeordneten Linien der Abbildung 3.

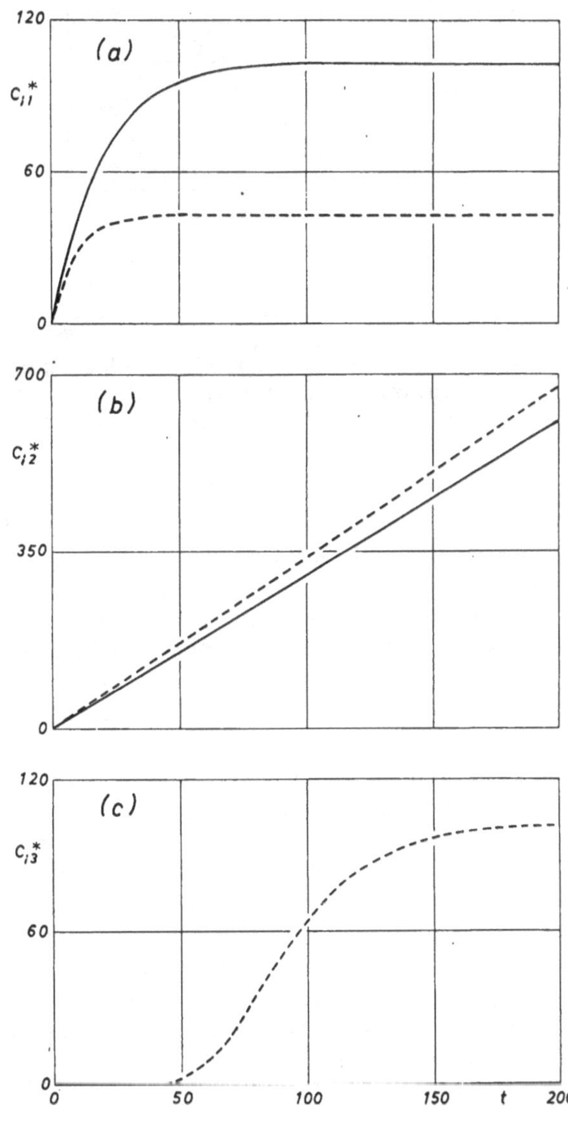

Abb. 4. Abhängigkeit der intrazellulären Ionenkonzentrationsänderungen $C_{i1}^*(a)$, $C_{i2}^*(b)$ und $C_{i3}^*(c)$ von der Versuchsdauer t. Gestrichelte Linien: Versuche mit Saccharoselösung; durchgezogene Linien: Versuche mit DNP-Saccharoselösung. Ordinaten: C_i^* in mMol/1000 cm³ Nervenfaser, Abszissen t in min

Den bisherigen Darlegungen zufolge veranschaulichen die Abbildungen 4 und 5, um welche Beträge die Chlor- und Kaliumkonzentrationen innerhalb der Nervenfasern während der Versuche durch den Ionenefflux abnehmen müßten. Eine Ausnahme bildet die durchgezogene Kurve der Abbildung 4a. Sie gibt den intrazellulären Konzentrationszuwachs der Chlorionen an, der durch Chloraufnahme bei Vergiftung der Präparate mit DNP zu erwarten wäre.

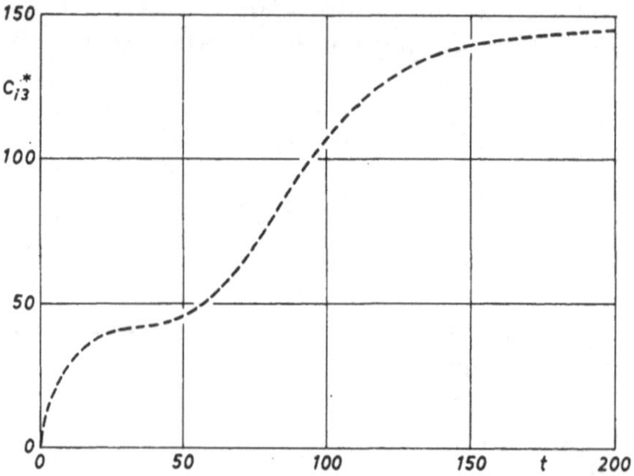

Abb. 5. Darstellung des biphasischen Verlaufs der Chlorabgabe aus den Nervenfasern. Ordinaten: Intrazelluläre Ionenkonzentrationsänderung $C_{i1}^* + C_{i3}^*$ in mMol/1000 cm³ Nervenfaser, Abszissen: Versuchsdauer t in min

Die intraaxonale Kaliumkonzentration der markhaltigen Nervenfaser wird von FRANKENHAEUSER [10] mit 120 mMol/1000 ml Faserwasser veranschlagt. Die Chlorkonzentration im Axon beträgt nach SHANES und BERMAN [32] 27 mMol, nach SCHMIDT [30] vielleicht sogar nur 7,43 mMol/1000 ml Faserwasser. Im Vergleich dazu liegen die Ordinatenwerte, die von den Kurven der Abbildungen 4 und 5 erreicht werden, viel zu hoch, denn die Präparate können nicht mehr Ionen abgeben, als zu Versuchsbeginn in ihnen enthalten waren. Selbst wenn man bedenkt, daß wahrscheinlich nicht nur die Achsenzylinder, sondern auch die Markscheiden der Nervenfasern am Ionenaustausch teilnehmen (vgl. SCHMIDT [30]), kann man die großen Konzentrationsänderungen kaum befriedigend erklären. Würden dagegen die „Chlor"- und „Kaliumfraktionen" außer Chlor- und Kaliumionen noch andere Stoffe enthalten, so ließen sich die Befunde zwanglos deuten, denn die Rechnung müßte dann für C_{i1}^*, C_{i2}^* und C_{i3}^* zu hohe Werte liefern. Wie schon erwähnt, ist auch aus Gründen der Elektroneutralität anzunehmen, daß die Präparate in den einzelnen Fraktionen nicht nur jeweils eine Ionensorte, sondern Ionengemische abgeben oder aufnehmen. Dabei könnten besonders mehrwertige organische Ionen eine Rolle spielen, weil von diesen Körpern wegen ihrer großen Ladungszahl nach Gleichung (15) schon geringe Mengen genügen würden, um den Leitwert der Testlösung und damit auch die aus der Formel (19) berechnete intrazelluläre Konzentrationsänderung stark zu vergrößern.

Es ist also nicht möglich, die Absolutwerte der intrazellulären Chlor- und Kaliumkonzentrationsänderungen durch Leitfähigkeitsmessungen in der geschilderten Weise zu ermitteln, aber die Gesetzmäßigkeiten, nach denen sich diese Konzentrationsänderungen vollziehen, können aus den Komponenten G_1^*, G_2^* und G_3^* oder aus C_{i1}^*, C_{i2}^* und C_{i3}^* hergeleitet werden. Nimmt man z. B. an, daß der unvergiftete Nerv in der ersten Fraktion außer Chlorionen noch die Ionensorten A, B, ..., Z abgibt, so resultiert für G_1^* die Gleichung

$$G_1^* = G_{Cl1}^* + G_{A1}^* + G_{B1}^* + \cdots + G_{Z1}^* \tag{22}$$

oder die Beziehung

$$G_1^* = \frac{n\frac{d^2}{4}\pi\lambda}{1000\, l^2}\{l_{Cl}\,[Cl]_{i1}^* + z_A\, l_A\, [A]_{i1}^* + z_B\, l_B\, [B]_{i1}^* + \cdots + z_Z\, l_Z\, [Z]_{i1}^*\}, \tag{23a}$$

wenn man nach der Formel (19) den Zusammenhang zwischen Leitwerts- und Konzentrationsänderung beachtet. Die eckigen Klammern deuten die Konzentrationen der betreffenden Stoffe in Mol/1000 cm³ Nervenfaser an. Multipliziert man die Summe (23a) mit dem Produkt (20a), so erhält man für C_{i1}^* die einfachere Gleichung

$$C_{i1}^* = \frac{1}{l_{Cl}}\{l_{Cl}\,[Cl]_{i1}^* + z_A\, l_A\, [A]_{i1}^* + z_B\, l_B\, [B]_{i1}^* + \cdots + z_Z\, l_Z\, [Z]_{i1}^*\}. \tag{23b}$$

Da sowohl G_1^* als auch C_{i1}^* dem Exponentialgesetz $(1 - e^{-a_g t})$ direkt proportional sind, muß diese Proportionalität auch für jedes Glied auf der rechten Seite der Gleichungen (23a) und (23b) gelten, denn eine Reihenentwicklung des Ausdruckes $(1 - e^{-a_g t})$ wäre nicht sinnvoll. Weil in den einzelnen Summanden aber nur die Faktoren $[Cl]_{i1}^*$, $[A]_{i1}^*$, $[B]_{i1}^*$, ..., $[Z]_{i1}^*$ von t abhängen, kann man schreiben

$$\left.\begin{aligned} G_1^* &\sim (1 - e^{-a_g t}) \\ C_{i1}^* &\sim (1 - e^{-a_g t}) \\ [Cl]_{i1}^* &\sim (1 - e^{-a_g t}) \\ [A]_{i1}^* &\sim (1 - e^{-a_g t})\,. \\ [B]_{i1}^* &\sim (1 - e^{-a_g t}) \\ &\cdots\cdots\cdots\cdots \\ [Z]_{i1}^* &\sim (1 - e^{-a_g t}) \end{aligned}\right\} \tag{24}$$

Aus der Zusammenstellung (24) folgt die Beziehung

$$G_1^* \sim C_{i1}^* \sim [Cl]_{i1}^*. \tag{25a}$$

Diese Überlegungen lassen sich auch auf die zweite und dritte Ionenfraktion anwenden, so daß man die Relationen

$$G_2^* \sim C_{i2}^* \sim [K]_{i2}^* \tag{25b}$$

und

$$G_3^* \sim C_{i3}^* \sim [Cl]_{i3}^* \tag{25c}$$

erhält. Mit Einführung der Proportionalitätsfaktoren α_1, α_2 und α_3 kann man nun die Gesetzmäßigkeiten, denen die Veränderungen der intrazellulären Chlor- und Kaliumkonzentrationen gehorchen, in Parallele zu den Ansätzen für die G^*- und C_i^*-Komponenten formulieren. Den Versuchen mit DNP-freier Saccharoselösung entsprechen die Beziehungen

$$[Cl]_{i1}^* = \alpha_1 (1 - e^{-a_g t}), \tag{26a}$$

$$[K]_{i2}^* = \alpha_2 t \tag{26b}$$

und

$$[Cl]_{i3}^* = \alpha_3 \int \tau^{c_g} e^{-b_g \tau} d\tau. \tag{26c}$$

Die daraus abgeleiteten Differentialgleichungen (vgl. BRADL [3], [4]) lauten

$$\frac{d^2 [\mathrm{Cl}]_{i1}^*}{dt^2} a_g \frac{d [\mathrm{Cl}]_{i1}^*}{dt} = 0 \,, \tag{27a}$$

$$\frac{d^2 [\mathrm{K}]_{i2}^*}{dt^2} = 0 \tag{27b}$$

und

$$\frac{d^2 [\mathrm{Cl}]_{i3}^*}{d\tau^2} + \left(b_g - \frac{c_g}{\tau}\right) \frac{d [\mathrm{Cl}]_{i3}^*}{d\tau} = 0 \,. \tag{27c}$$

Bei Verwendung DNP-haltiger Saccharoselösung gelten nur die Ansätze (26a) und (26b), sowie die Differentialgleichungen (27a) und (27b). Die Faktoren α_1 und α_2 können hier andere Beträge haben als bei unvergifteten Präparaten. Die Veränderungen der Chlor- und Kaliumkonzentrationen im Inneren der Nervenfasern unterliegen somit denselben Gesetzen wie die Veränderungen der elektrischen Leitfähigkeit des äußeren Mediums.

2. *Berechnung der Ionenfluxe*

Wenn eine Ionensorte x zwischen den Nervenfasern und dem umgebenden Medium ausgetauscht wird, so bezeichnet man die Ionenmenge, die dabei pro Zeiteinheit durch die Einheit der Zelloberfläche hindurchtritt, als Ionenflux. Diese Größe wird meist in pMol/cm² sec angegeben. Findet der Stofftransport nur vom Faserinneren zum Extrazellulärraum hin statt, so ist er mit einem Efflux M_{xa} verknüpft; strömen die Ionen lediglich in entgegengesetzter Richtung, so resultiert ein Influx M_{xi}. Da meist beide Vorgänge gleichzeitig ablaufen, kann man einen Netto-Flux M_x nach der Gleichung

$$M_x = M_{xa} - M_{xi} \tag{28}$$

definieren. Wenn M_{xa} größer als M_{xi} ist, stellt M_x einen Netto-Efflux, im umgekehrten Fall einen Netto-Influx dar.

Die Kenntnis der Fluxbeziehungen ist für das Studium des Stofftransportes zwischen dem intra- und extrazellulären Milieu unerläßlich. Deshalb soll gezeigt werden, welche Schlußfolgerungen für die Chlor- und Kaliumfluxe bei markhaltigen Nervenfasern aus den bisherigen Darlegungen zu ziehen sind.

Wie die intrazelluläre Konzentrationsänderung C_{ix}^* läßt sich auch M_x aus den Ergebnissen der Leitfähigkeitsmessungen ableiten. Eine gesonderte Erfassung von M_{xi} und M_{xa} ist dabei nicht möglich, weil sich beide Größen überlagern und somit nur die Differenz (28) in G_x^* ihren Ausdruck finden kann.

Bei den Versuchen wurde die elektrische Leitfähigkeit der Testlösungen in Abständen von 10 min registriert. Bezeichnet man die Leitwertsänderung in einem solchen Zeitabschnitt mit ΔG^*, so folgt aus Gleichung (17) der Ansatz

$$|\Delta C_x^*| = \frac{l^2}{z_x l_x} \Delta G_x^* . \qquad (29)$$

Die absolute Ionenmenge $|\delta C_x^*|$, die in einer Sekunde vom Nerven abgegeben oder aufgenommen wird, resultiert, wenn man das Produkt (29) durch 600 dividiert:

$$|\delta C_x^*| = \frac{l^2}{z_x l_x} \frac{1}{600} \Delta G_x^* . \qquad (30)$$

Aus dieser Formel läßt sich M_x ermitteln, sofern die Oberfläche O_N aller im Präparat enthaltener Nervenfasern bekannt ist. Für O_N gilt die Beziehung

$$O_N = n\, d\, \pi\, \lambda. \qquad (31)$$

Bei Division der Größe $|\delta C_x^*|$ durch O_N ergibt sich M_x zu

$$M_x = \frac{l^2}{z_x l_x} \frac{1}{600 n d \pi \lambda} \Delta G_x^* . \qquad (32)$$

Nach dieser Gleichung wurden die Fluxe M_1, M_2 und M_3 aus den Leitwertskomponenten G_1^*, G_2^* und G_3^* bestimmt. Wie bei der Berechnung von C_{i1}^*, C_{i2}^* und C_{i3}^* habe ich für die Ionenbeweglichkeiten bei M_1 und M_3 die Konstante l_{Cl}, bei M_2 die Größe l_K eingesetzt. Die

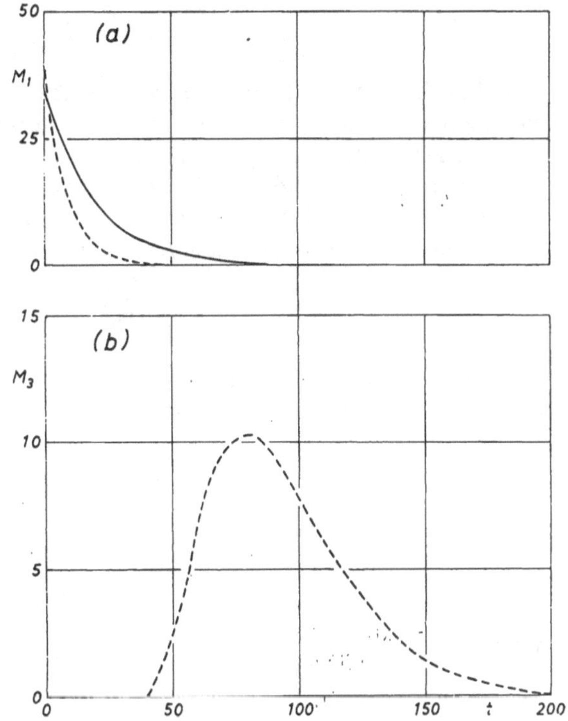

Abb. 6. Abhängigkeit der Ionenfluxe $M_1(a)$ und $M_3(b)$ von der Versuchsdauer t. Gestrichelte Linien: Versuche mit Saccharoselösung; durchgezogene Linie: Versuche mit DNP-Saccharoselösung. Ordinaten: M in pMol/cm² sec, Abszisse: t in min

erhaltenen M_1- und M_3- Werte sind in pMol/cm² sec in Abbildung 6 als Funktionen der Versuchsdauer t dargestellt. Die gestrichelten Kurven geben die Netto-Effluxe bei unvergifteten Nervenfasern wieder, die durchgezogene Linie veranschaulicht den Verlauf des Netto-Influxes, der bei Behandlung der Präparate mit DNP entsteht.
Die Größen M_1 werden exponentiell mit der Zeit kleiner, wobei der Efflux rascher versiegt als der Influx. M_3 beginnt bei $t = 40$ min mit einem steilen Anstieg, der bei $t = 80$ min in einen langsameren Abfall übergeht. M_2 bleibt während der ganzen Versuchsdauer konstant und beträgt beim normalen Faserbündel 20,833 pMol/cm² sec, bei DNP-Vergiftung 19,318 pMol/cm² sec. Die Differenz zwischen diesen Werten ist sehr gering, in beiden Fällen liegt ein Netto-Efflux vor.
Den bisherigen Ausführungen zufolge sind die Fluxe den zeitlichen Änderungen der entsprechenden Leitwertskomponenten proportional. M_1, M_2 und M_3 müssen sich also durch die ersten Ableitungen von G_1^*, G_2^* und G_3^* nach der Zeit beschreiben lassen. Es gelten die Beziehungen

$$M_1 = h_{Cl} \frac{dG_1^*}{dt}, \qquad (33\,a)$$

$$M_2 = h_K \frac{dG_2^*}{dt} \qquad (33\,b)$$

und

$$M_3 = h_{Cl} \frac{dG_3^*}{dt} \qquad (33\,c)$$

mit den Proportionalitätsfaktoren h_{Cl} und h_K. Um h_{Cl} und h_K zu ermitteln, habe ich als Beispiel die Größe dG_1^*/dt einmal aus der Funktion (12) gebildet, zum anderen durch graphisches Differenzieren der Kurven in Abbildung 3a gewonnen. Ein Vergleich der Resultate hat gelehrt, daß die Differentialquotienten mit völlig hinreichender Genauigkeit durch Differenzenquotienten gemäß der verallgemeinerten Bedingung

$$\frac{dG_x^*}{dt} = \frac{\Delta G_x^*}{10} \qquad (34)$$

ersetzt werden können. Damit ergeben sich für die Konstanten h_{Cl} und h_K aus Gleichung (32) die Produkte

$$h_{Cl} = \frac{l^2}{l_{Cl}} \frac{1}{60\, n\, d\, \pi\, \lambda} \qquad (35\,a)$$

und

$$h_K = \frac{l^2}{l_K} \frac{1}{60\, n\, d\, \pi\, \lambda}. \qquad (35\,b)$$

Werden die Ableitungen in den Formeln (33a), (33b) und (33c) unter Beachtung der Ansätze (12), (9) und (10) ausgeschrieben, so erhält man Funktionen, die den Zeitgang der Fluxkomponenten darstellen. Für unvergiftete

Nervenfasern gelten die Beziehungen

$$M_1 = h_{Cl}\, a_g\, \chi\, e^{-a_g t}, \tag{36a}$$

$$M_2 = h_K\, \psi \tag{36b}$$

und

$$M_3 = h_{Cl}\, \omega\, \tau^{c_g}\, e^{-b_g \tau}. \tag{36c}$$

Bei Vergiftung der Präparate mit DNP resultieren lediglich die Gleichungen (36a) und (36b).
Wie aus den Konzentrationsänderungen C_{i1}^*, C_{i2}^* und C_{i3}^* die Funktionen $[Cl]_{i1}^*$, $[K]_{i2}^*$ und $[Cl]_{i3}^*$ berechnet worden sind, so lassen sich nach demselben Prinzip auch die Fluxe der Chlor- und Kaliumionen M_{Cl1}, M_{K2} und M_{Cl3} aus den Größen M_1, M_2 und M_3 herleiten. Man kann also in Analogie zu den Proportionen (25a), (25b) und (25c) die Zusammenhänge

$$\Delta G_1^* \sim M_1 \sim M_{Cl1}, \tag{37a}$$

$$\Delta G_2^* \sim M_2 \sim M_{K2} \tag{37b}$$

und

$$\Delta G_3^* \sim M_3 \sim M_{Cl3} \tag{37c}$$

aufstellen. Mit Einführung der Proportionalitätsfaktoren β_1, β_2 und β_3 ergibt sich danach die Abhängigkeit der Fluxe M_{Cl1}, M_{K2} und M_{Cl3} von der Versuchsdauer aus den Formeln (36a), (36b) und (36c). Die entsprechenden Ausdrücke lauten für unvergiftete Nervenfaserbündel

$$M_{Cl1} = \beta_1\, e^{-a_g t}, \tag{38a}$$

$$M_{K2} = \beta_2 \tag{38b}$$

und

$$M_{Cl3} = \beta_3\, \tau^{c_g}\, e^{-b_g \tau}. \tag{38c}$$

Dabei gelten die Differentialgleichungen

$$\frac{dM_{Cl1}}{dt} + a_g M_{Cl1} = 0, \tag{39a}$$

$$\frac{dM_{K2}}{dt} = 0 \tag{39b}$$

und

$$\frac{dM_{Cl3}}{d\tau} + \left(b_g - \frac{c_g}{\tau}\right) M_{Cl3} = 0\,. \tag{39c}$$

Vergiftet man die Präparate mit DNP, so ergeben sich die Ansätze (38a) und (38b) mit den Differentialgleichungen (39a) und (39b). Die Faktoren β_1 und β_2 können sich hierbei betragsmäßig von den analogen Konstanten der nicht vergifteten Präparate unterscheiden. Dabei ist zu beachten, daß nur M_{Cl1} beim DNP-vergifteten Nerven einen Netto-Influx beschreibt, alle übrigen Funktionen stellen Netto-Effluxe dar.

3. *Fluxgröße und intrazelluläre Ionenkonzentration*

Nachdem die Abhängigkeit der Ionenfluxe von der Versuchsdauer beschrieben worden ist, interessiert nun die Frage, welcher Zusammenhang zwischen den Fluxen und den Konzentrationsänderungen besteht, die im Inneren der Nervenfasern durch das Aus- oder Einströmen von Ionen verursacht werden. In Abbildung 7 sind deshalb die Größen M_1 und M_3 als Funktionen der zugehörigen C_i^*-Werte dargestellt. Die gestrichelten Linien geben die Efflux-Verhältnisse bei unvergifteten Präparaten wieder, die durchgezogene Gerade veranschaulicht den entsprechenden Influx-Verlauf bei Behandlung der Faserbündel mit DNP.

Wie Abbildung 7a zeigt, wird M_1 in beiden Fällen mit wachsendem C_{i1}^* kleiner. Die Verknüpfung von M_1 und C_{i1}^* ist linear. Beim normalen Nerven geht also die durch den Ionenverlust bedingte Verringerung des intrazellulären Ionengehaltes mit einer zunehmenden Drosselung der ionalen Effluxrate einher. Unter DNP-Einfluß wird analog der Influx vermindert, wenn die Ionenkonzentration im Faserinneren durch Ionenaufnahme ansteigt. Die unterschiedliche Steilheit der Geraden läßt erkennen, daß dieser Effekt beim Efflux viel deutlicher in Erscheinung tritt als beim Influx.

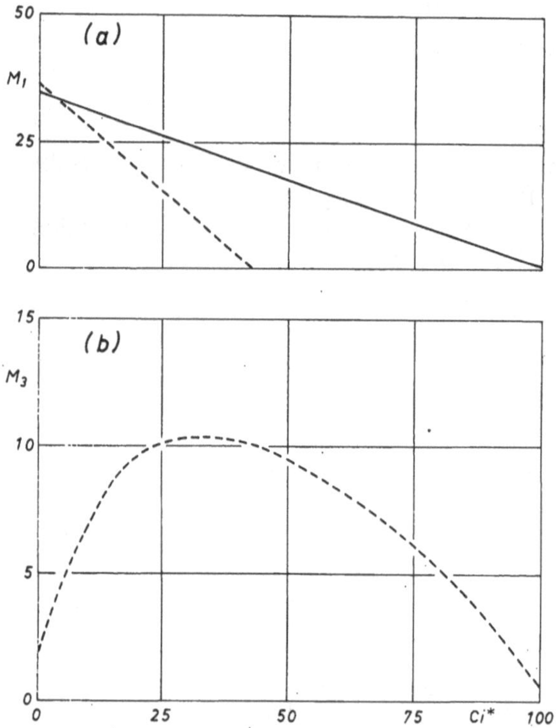

Abb. 7. Abhängigkeit der Ionenfluxe $M_1(a)$ und $M_3(b)$ von der zugehörigen intrazellulären Ionenkonzentrationsänderung C_i^*.

Gestrichelte Linien: Versuche mit Saccharoselösung; durchgezogene Linie: Versuche mit DNP-Saccharoselösung. Ordinaten: M in pMol/cm² sec, Abszissen: C_i^* in mMol/1000 cm³ Nervenfaser

Die Beziehungen, die nach Abbildung 7a zwischen M_1 und C_{i1}^* bestehen, können auch auf der Grundlage der Formeln (21a) und (36a) abgeleitet werden. Aus dem Ansatz (21a) läßt sich der Faktor

$$e^{-a_g t} = 1 - \frac{C_{i1}^*}{k_{Cl}\chi} \qquad (40)$$

eliminieren und in den Ausdruck (36a) einsetzen. Damit erhält man die Gleichung

$$M_1 = h_{Cl} a_g \chi - \frac{h_{Cl} a_g}{k_{Cl}} C_{i1}^*, \qquad (41)$$

die einer fallenden Geraden entspricht. Auf dieselbe Weise kann man aus den Funktionen

(26a) und (38a) den Zusammenhang zwischen M_{Cl1} und $[Cl]^*_{i1}$ ermitteln. Es resultiert die Formel

$$M_{Cl1} = \beta_1 - \frac{\beta_1}{\alpha_1} [Cl]^*_{i1} . \tag{42}$$

Auch diese Gleichung beschreibt eine fallende Gerade. Die Fluxe der Chlorionen verändern sich also mit der intrazellulären Konzentration dieser Ionen im Prinzip ebenso, wie M_1 mit C^*_{i1}. Ein solches Verhalten der Flux-Konzentrations-Beziehungen ist nach den Proportionen (25a), (25b), (25c), (37a), (37b) und (37c) auch allgemein zu erwarten.

Wesentlich komplizierter als in Abbildung 7a liegen die Verhältnisse in Abbildung 7b. Trotz Verminderung der Ionenkonzentration in den Nervenfasern wird der Efflux M_3 zunächst größer, um erst bei stärkerem Ionenverlust der Präparate wieder abzunehmen.

Auch zwischen M_2 und C^*_{i2} besteht anscheinend kein einfacher Zusammenhang, da der Efflux hier während der ganzen Versuchsdauer konstant bleibt, obwohl die Faserbündel laufend Ionen abgeben.

Die Fluxkomponenten M_1, M_2 und M_3 und damit auch M_{Cl1}, M_{K2} und M_{Cl3} unterscheiden sich also nicht nur als Funktionen der Zeit, sondern auch in ihren Beziehungen zur intrazellulären Ionenkonzentration sehr wesentlich voneinander.

4. *Passiver und aktiver Ionentransport*

Bekanntlich kommen für den Ionenaustausch zwischen der Zelle und ihrem umgebenden Medium sowohl passive als auch aktive Vorgänge in Betracht. Wenn Ionenverschiebungen stattfinden, ohne daß hierzu Energie aus dem Stoffwechsel benötigt wird, so liegt ein passiver Transportmechanismus vor. Im einfachsten Fall handelt es sich dabei um einen Diffusionsvorgang, bei dem Ionen durch die entsprechend permeable Zellmembran hindurchtreten. Es gibt auch Ionenbewegungen, die einen Transportvermittler benötigen, aber keine Stoffwechselenergie beanspruchen (NETTER [25]). Diese Prozesse sind gleichfalls passiv. Im Gegensatz dazu ist eine aktive Beförderung von Ionen zwischen dem intra- und extrazellulären Raum mit Stoffwechselleistungen der Zelle verknüpft. Die Energie für den aktiven Ionentransport wird häufig von oxidativen Phosphorylierungsprozessen geliefert, denn durch Ausschaltung dieser chemischen Reaktionen mit DNP lassen sich viele Transportvorgänge sehr wirksam blockieren (vgl. z. B. HODGKIN und KEYNES [17], [19], CALDWELL et al. [6], [7], KEYNES [22], KLEINZELLER und KOTYK [23], NETTER [25] und PASSOW [26]).

Wie die vorliegenden Befunde zeigen, sind oxidative Phosphorylierungen auch für den Ionentransport bei ruhenden markhaltigen Nervenfasern von Bedeutung. Der Chloraustausch zwischen den Faserbündeln und dem Außenmilieu wird durch DNP beeinflußt, wobei der Netto-Efflux der Chlorionen abnimmt und einem Netto-Influx Platz macht. Der Kaliumefflux hängt dagegen kaum mit diesen Stoffwechselvorgängen zusammen, weil er durch DNP nicht merklich verändert wird. Orientierende Experimente, bei denen die Präparate mit Natriumzyanid und Monojodessigsäure vergiftet worden sind, haben keinen Anhaltspunkt dafür geliefert, daß der Kaliumtransport Energie aus sonstigen oxidativen Prozessen oder aus dem Glykolysestoffwechsel benötigt. Die Kaliumabgabe aus markhaltigen Nervenfasern erfolgt also wahrscheinlich passiv.

Es erhebt sich die Frage, ob M_{K2} auf einem einfachen Diffusionsvorgang der Kaliumionen ohne Transportvermittler beruhen kann. Ein solcher Prozeß ist nur dann möglich, wenn das elektrochemische Potential $\bar{\mu}_K$ dieser Ionen im Zellinneren einen anderen Betrag hat als im Extrazellulärraum (vgl. KORTÜM [24]). Die treibende Kraft für den resultierenden Ionenflux M_K ist der Gradient $d\bar{\mu}_K/dm$ dieses Potentials, wobei m die Dicke der Zellmembran darstellt. M_K ist dem Potentialgradienten proportional:

$$M_K \sim \frac{d\bar{\mu}_K}{dm}. \tag{43}$$

Wenn $\bar{\mu}_K$ innerhalb der Membran linear abfällt, gilt der Zusammenhang

$$\frac{d\bar{\mu}_K}{dm} = \text{const} = \frac{\bar{\mu}_{Ki} - \bar{\mu}_{Ka}}{m} = \frac{\Delta\bar{\mu}_K}{m}, \tag{44}$$

wobei die Indizes i und a die Größe $\bar{\mu}_K$ jeweils auf die innere oder auf die äußere Seite der Membran beziehen und $\Delta\bar{\mu}_K$ die Potentialdifferenz zwischen innen und außen angibt. Aus den Gleichungen (43) und (44) folgt, daß auch M_K und $\Delta\bar{\mu}_K$ einander proportional sind, was man durch die Formel

$$M_K = \frac{g_K}{F^2}\,\Delta\bar{\mu}_K \tag{45}$$

ausdrücken kann. Hierin bedeuten g_K die Kaliumleitfähigkeit der Membran und F die FARADAY-Konstante (vgl. HODGKIN und KEYNES [18] sowie KATCHALSKY [21]). Durch den Ionenfluß wird ein Gleichgewichtszustand angestrebt, für den nach KORTÜM [24] die Bedingung

$$\bar{\mu}_{Ki} = \bar{\mu}_{Ka} \tag{46}$$

oder gemäß der Beziehung (44)

$$\Delta\bar{\mu}_K = 0 \tag{47}$$

gilt. $\Delta\bar{\mu}_K$ verkleinert sich also während des Diffusionsvorganges. Die Abnahme von $\Delta\bar{\mu}_K$ pro Zeiteinheit ist hierbei dem augenblicklichen Wert von $\Delta\bar{\mu}_K$ jeweils proportional. Deshalb kann man schreiben

$$-\frac{d\Delta\bar{\mu}_K}{dt} = \gamma\,\Delta\bar{\mu}_K. \tag{48}$$

γ stellt eine Konstante dar. Die Lösung der Differentialgleichung (48) liefert den Exponentialausdruck

$$\Delta\bar{\mu}_K = \Delta\bar{\mu}_{K0}\,e^{-\gamma t}, \tag{49}$$

worin $\Delta\bar{\mu}_{K_0}$ die elektrochemische Potentialdifferenz bedeutet, die zu Beginn des Diffusionsprozesses an der Nervenmembran herrscht. Aus den Formeln (45) und (49) erhält man für M_K den Zusammenhang

$$M_K = \frac{g_K}{F^2} \Delta\bar{\mu}_{K_0} e^{-\gamma t}. \tag{50}$$

Die Gleichung (38b) zeigt, daß der Kaliumefflux während der Versuchsdauer konstant bleibt; nach dem Ansatz (50) müßte er dagegen im Laufe der Zeit kleiner werden. Dieser Widerspruch läßt sich beseitigen, wenn man bedenkt, daß g_K keine invariable Größe ist, sondern vom Membranpotential abhängt (Hodgkin und Huxley [13], [14], Stämpfli [34], Frankenhaeuser [10]). Werden die Nervenfasern um den Betrag ΔU depolarisiert, so gilt in gewissen Grenzen die Beziehung

$$g_K = g_{K_0} e^{\frac{\Delta U}{\varkappa}}. \tag{51}$$

Hierin stellen g_{K_0} den Membranleitwert für Kaliumionen beim normalen Ruhepotential und $\varkappa$ eine Konstante dar. Nun ist der Kaliumefflux sehr wahrscheinlich mit einem Depolarisationsprozeß verbunden, der linear mit der Zeit zunimmt (Bradl [3], [4]). Das kann man durch die Formel

$$\Delta U = \zeta t \tag{52}$$

ausdrücken, wobei ζ ein Proportionalitätsfaktor ist. Aus den Funktionen (51) und (52) resultiert dann der Zusammenhang

$$g_K = g_{K_0} e^{\frac{\zeta}{\varkappa} t}, \tag{53}$$

womit die Gleichung (50) in den Ausdruck

$$M_K = \frac{\Delta\bar{\mu}_{K_0}}{F^2} g_{K_0} e^{\left(\frac{\zeta}{\varkappa} - \gamma\right) t} \tag{54}$$

übergeht. Wenn dabei noch die Forderung

$$\frac{\zeta}{\varkappa} = \gamma \tag{55}$$

erfüllt ist, erhält man gemäß der Beziehung (54) in Übereinstimmung mit der Formel (38b) das Ergebnis

$$M_K = \frac{\Delta\bar{\mu}_{K_0}}{F^2} g_{K_0} = \text{const}. \tag{56}$$

M_{K2} kann also durchaus auf einem einfachen Diffusionsprozeß der Kaliumionen beruhen. Die Annahme eines besonderen Transportvermittlers ist nicht erforderlich.

Selbstverständlich wird der Ionenflux nicht ad infinitum konstant bleiben. Das ist schon aus thermodynamischen Gründen unmöglich, denn die Gleichgewichtsbedingung (46) muß irgendwann einmal erreicht werden. Wenn der Kaliumverlust der Nervenfasern und damit ΔU einen bestimmten Betrag überschreiten, ändert sich g_K mit weiter zunehmender Depolarisation nicht mehr wesentlich (Hodgkin und Huxley [13] sowie Frankenhaeuser [10]). Die Formel (51) und alle daraus abgeleiteten Beziehungen verlieren dann ihre Gültigkeit und M_K erlischt gemäß Gleichung (50). Nach den eigenen Befunden treten diese Vorgänge aber offenbar erst ein, wenn die Versuche bedeutend länger als 200 min dauern.

Etwas schwieriger als die Kaliumabgabe aus den Nervenfasern ist der Austausch der Chlorionen zu beurteilen. Vergleicht man die Ansätze (38a) und (50) miteinander, so fällt auf, daß diese Fluxe im Prinzip denselben Zeitgang haben. Kinetisch verhält sich M_{Cl1} also wie ein passiver Diffusionsvorgang, während andererseits DNP die Flußrichtung umkehrt und die Zeitkonstante $1/a_g$ dieses Prozesses vergrößert (vgl. Tab. 2). Nun besteht kein Grund zu der Hypothese, daß der Chlorgehalt der Testlösung durch den DNP-Zusatz verändert wird. Der Unterschied der Chlorionenkonzentration zwischen intra- und extrazellulärem Raum ist also zu Versuchsbeginn beim normalen Faserbündel derselbe, wie bei den Experimenten mit DNP. Deshalb läßt sich der Richtungswechsel von M_{Cl1} auf der Basis einer einfachen Diffusion nicht deuten; vielmehr ist anzunehmen, daß die Chlorionen in einem von beiden Fällen entgegen ihrem Konzentrationsgradienten befördert werden, was ein Kennzeichen von aktiven Transportvorgängen ist (Netter [25] und Ussing [36]).

Bei M_{Cl3} spricht schon die Tatsache, daß dieser Efflux durch DNP völlig unterdrückt werden kann, gegen ein passives Geschehen. Hinzu kommt noch der verwickelte zeitliche Verlauf der Funktion (38c), der gleichfalls den Gedanken an einen komplizierten Transportvorgang nahelegt. Wollte man nämlich M_{Cl3} wie M_{K2} auf einen einfachen Diffusionsprozeß der Chlorionen zurückführen, so müßte man für g_{Cl} die Forderung

$$g_{Cl} \sim \tau^{c_0} \tag{57}$$

stellen, um die Gleichungen (38c) und (50) miteinander in Einklang zu bringen. Gemäß der Definition (13) dürfte dann eine Chlorleitfähigkeit der Nervenmembran normalerweise gar nicht bestehen, sondern sie würde erst durch längere Saccharosebehandlung der Faserbündel erzeugt und müßte außerdem parabolisch mit der Zeit größer werden. Ein solches Verhalten erscheint aber sehr abwegig.

Der Efflux der Chlorionen aus den Nervenfasern ist also wahrscheinlich an einen aktiven Transportmechanismus gebunden, der Energie aus dem oxidativen Phosphatstoffwechsel bezieht. Dabei können die Chlorionen entweder selbst durch einen Carrier befördert werden oder sie wandern auf Grund eines elektrischen Feldes, das durch den aktiven Transport einer anderen Ionensorte in der Zellmembran entsteht (Ussing [36]).

Anschrift des Verfassers: Dr. med. M. Bradl, Physiologisches Institut der Friedrich-Schiller-Universität Jena, Teichgraben 8.

Literatur

[1] Böhm, H. W. u. R. W. Straub: Pflügers Arch. ges. Physiol. Menschen Tiere **274**, 468 (1962).
[2] Bradl, M.: Acta biol. med. german. **6**, 467 (1961).
[3] Bradl, M.: Acta biol. med. german. **9**, 247 (1962).
[4] Bradl, M.: Acta biol. med. german. **16**, 154 (1966).

[5] BRONSTEIN, I. N. u. K. A. SEMENDJAJEW: Taschenbuch der Mathematik. Leipzig 1958, B. G. Teubner Verlagsgesellschaft.
[6] CALDWELL, P. C., A. L. HODGKIN, R. D. KEYNES u. T. I. SHAW: J. Physiology **152**, 561 (1960).
[7] CALDWELL, P. C., A. L. HODGKIN, R. D. KEYNES u. T. I. SHAW: J. Physiology **152**, 591 (1960).
[8] D'ANS, J. u. E. LAX: Taschenbuch für Chemiker und Physiker. Berlin/Göttingen/Heidelberg 1949, Springer-Verlag.
[9] DETTBARN, W. D. u. R. STÄMPFLI: Helvet. physiol. pharmacol. Acta **15**, 25 (1957).
[10] FRANKENHAEUSER, B.: J. Physiology **160**, 40 (1962).
[11] HODGKIN, A. L.: Biol. Rev. **26**, 339 (1951).
[12] HODGKIN, A. L. u. A. F. HUXLEY: Nature [London] **144**, 710 (1939).
[13] HODGKIN, A. L. u. A. F. HUXLEY: J. Physiology **116**, 449 (1952).
[14] HODGKIN, A. L. u. A. F. HUXLEY: J. Physiology **117**, 500 (1952).
[15] HODGKIN, A. L., A. F. HUXLEY u. B. KATZ: Arch. Sci. physiol. **3**, 129 (1949).
[16] HODGKIN, A. L. u. B. KATZ: J. Physiology **108**, 37 (1949).
[17] HODGKIN, A. L. u. R. D. KEYNES: J. Physiology **128**, 28 (1955).
[18] HODGKIN, A. L. u. R. D. KEYNES: J. Physiology **128**, 61 (1955).
[19] HODGKIN, A. L. u. R. D. KEYNES: J. Physiology **131**, 592 (1956).
[20] KARAOGLANOW, Z.: Fresenius' Z. analyt. Chem. **115**, 305 (1939).
[21] KATCHALSKY, A.: In: A. KLEINZELLER u. A. KOTYK: Membrane Transport and Metabolism. Proceedings of a Symposium. Prag 1960, Publishing House of the Czechoslovak Academy of Sciences.
[22] KEYNES, R. D.: In: Biochemie des aktiven Transports. 12. Colloqu. d. Ges. f. Physiol. Chem., Mosbach/Baden, Berlin/Göttingen/Heidelberg 1961, Springer-Verlag.
[23] KLEINZELLER, A. u. A. KOTYK: Membrane Transport and Metabolism. Proceedings of a Symposium. Prag 1960, Publishing House of the Czechoslovak Academy of Sciences.
[24] KORTÜM, G.: Lehrbuch der Elektrochemie. 2. Aufl. Weinheim/Bergstr. 1957, Verlag Chemie G.m.b.H.
[25] NETTER, H.: In: Biochemie des aktiven Transports. 12. Colloqu. d. Ges. f. Physiol. Chem., Mosbach/Baden, Berlin/Göttingen/Heidelberg 1961, Springer-Verlag.
[26] PASSOW, H.: In: Biochemie des aktiven Transports. 12. Colloqu. d. Ges. f. Physiol. Chem., Mosbach/Baden, Berlin/Göttingen/Heidelberg 1961, Springer-Verlag.
[27] RITCHIE, J. M. u. R. W. STRAUB: J. Physiology **136**, 80 (1957).
[28] RUD, J.: Acta physiol. scand., Suppl. **51**, **178**, 95 (1961).
[29] SCHMIDT, H.: Pflügers Arch. ges. Physiol. Menschen Tiere **274**, 632 (1962).
[30] SCHMIDT, H.: Ann. Univ. sarav. Med. **11**, 2 (1964).
[31] SCHOEPFLE, G. M. u. F. E. BLOOM: Amer. J. Physiol. **197**, 1131 (1959).
[32] SHANES, A. M. u. M. D. BERMAN: J. cellular comparat. Physiol. **45**, 199 (1955).
[33] STÄMPFLI, R.: Ergebn. Physiol., biol. Chem. exp. Pharmacol. **47**, 70 (1952).
[34] STÄMPFLI, R.: Helvet. physiol. pharmacol. Acta **16**, 127 (1958).
[35] STRAUB, R.: Helvet. physiol. pharmacol. Acta **14**, 1 (1956).
[36] USSING, H. H.: In: Biochemie des aktiven Transports. 12. Colloqu. d. Ges. f. Physiol. Chem., Mosbach/Baden, Berlin/Göttingen/Heidelberg 1961, Springer-Verlag.

Summary

BRADL M.: Transport of ions in myelinated nerve fibres

The exchange of chloride and potassium ions between myelinated nerve fibres and the surrounding fluid (isotonic sucrose solution) has been studied. In some cases the fibres were poisened by 2,4-dinitrophenol.

The time courses of the ionic net fluxes and of the alteration of the intracellular ionic concentration have been calculated and described by differential equations.
From unpoisened nerve fibres the chloride ions escape in two different fractions, the potassium ions however in one fraction. Treatment by dinitrophenol diminishes the efflux and promotes the influx of chloride ions but does not alter the potassium efflux remarkably.
The results lead to the assumption, that the chloride efflux is connected with active transport basing on oxidative phosphorylation, while the potassium efflux can be explained by a simple passive diffusion process.

Резюме

М. Брадль: Ионный обмен в мозгосодержащих нервных волокнах

Измерением проводимости, химическими анализами и исследованиями структуры изучается обмен ионов хлора и калия между пучками мозгосодержащих нервных волокон в изотоническом растворе сахарозы с добавлением и без добавления 2,4-динитрофенола.
Для соответствующих изменений интрацеллюлярной концентрации ионов, а также для иональных потоков нетто выводятся дифференциальные уравнения, характеризующие изменение этих параметров во времени.
Неотравленные нервные волокна отдают ионы хлора в двух фракциях, а ионы калия — в одной фракции. При отравлении препаратов динитрофенолом вторая фракция хлора исчезает, в то время как в первой эффлюкс превращается в инфлюкс. Отдача калия не изменяется при добавлении динитрофенола.
Предполагается, что эффлюкс хлора обусловлен активным механизмом транспорта с энергетической связью с окислительным фосфатным обменом, в то время как эффлюкс калия представляет собой пассивный процесс диффузии.

Acta biol. med. german., Band 17, Seite 87—95 (1966)

Aus dem Physiologischen Institut der Friedrich-Schiller-Universität Jena
(Direktor: Prof. Dr. F. Schwarz)

Über die Wirkung einer Amidopyrin-Phenylbutazon-Kombination (*Wofapyrin*) auf den Ranvierschen Schnürring des Kaltblüternerven[1]

F. Schwarz und R. Thiele

(Eingegangen am 4. 3. 1966)

Zusammenfassung

Ranviersche Schnürringe motorischer Nervenfasern aus dem Nervus ischiadicus des Frosches werden mit rechteckigen elektrischen Wechselimpulsen gereizt. Unterschwellige lokale Potentiale und Aktionspotentiale werden abgeleitet. Die einzelnen Reizimpulse bringen lokale Potentialänderungen hervor, die sich summieren. Der Schnürring wird mit Wofapyrinlösung bespült.

Die Reizschwelle des einzelnen Wechselimpulses steigt unter Wofapyrinwirkung steil an; diese Wirkung ist zum Teil auswaschbar.

Die Schwellenspannungen für Impulsgruppen einer Frequenz von 2 kHz liegen niedriger als die für die Einzelimpulse; dies ist eine Folge der Summation. Die Summation ist unter Wofapyrinwirkung stärker als am normal bespülten Schnürring.

Bei Schwellenreizspannung wird die Zahl der Impulse registriert, nach deren Ablauf es zum Ausklinken eines Aktionspotentials kommt. Unter Wofapyrinwirkung ist die Impulszahl größer.

Wofapyrin setzt die Erregbarkeit des Schnürrings herab, vielleicht durch eine Verfestigung der erregbaren Membran.

Wofapyrin, ein zusammengesetztes Präparat (Dimethylaminophenyldimethylpyrazolon + 3,5-Dioxo-1,2-diphenyl-4-n-butylpyrazolidin), ist ein Antipyretikum und Antineuralgikum, das auch in der Therapie des Rheumatismus mit Erfolg angewendet wird. Über seinen Angriffsort und seine Wirkungsweise ist noch wenig bekannt. In der vorliegenden Mitteilung soll über Versuche berichtet werden, die an der motorischen Einzelfaser aus dem Nervus ischiadicus des Frosches angestellt wurden. Unterschwellige elektrische Wechselimpulse bringen am Ranvierschen Schnürring Änderungen des lokalen Potentials und Summationserscheinungen hervor (Schwarz und Volkmer [1], Schwarz [2]). Es fragte sich, wie Wofapyrin die Reizwirkung der Wechselimpulse beeinflußt.

Methodik

Zur Gewinnung eines Einzelfaserpräparates wurde ein in der üblichen Weise hergestelltes Nerv-Muskelpräparat des Nervus ischiadicus von Rana temporaria oder Rana

[1] Die Untersuchungen wurden im Rahmen des Zentrums der Rheumaforschung an der Medizinischen Fakultät der Friedrich-Schiller-Universität Jena durchgeführt

esculenta 20 min in gekühlter TASAKI-Lösung belassen. Dann erfolgte die Präparation unter einem Binokular-Präparier-Mikroskop bei Dunkelfeldbeleuchtung nach STÄMPFLI [3]. Der Nerv wurde zunächst von begleitenden Blutgefäßen und von Bindegewebe befreit, dann wurde das Epineurium des Nervus tibialis entfernt. Der Nervus peronaeus und die Äste des Nervus tibialis mit Ausnahme des oberen Astes für den Musculus gastrocnemius wurden abgeschnitten. Aus dem oberen Ast wurde dann eine motorische Faser isoliert. Die Reste der zerstörten Fasern wurden mit dem Messer beseitigt. Das Präparat wurde in eine Versuchsanordnung gebracht, wie sie STÄMPFLI [4], KILB und STÄMPFLI [5] und RUDOLPH und STÄMPFLI [6] angegeben haben.

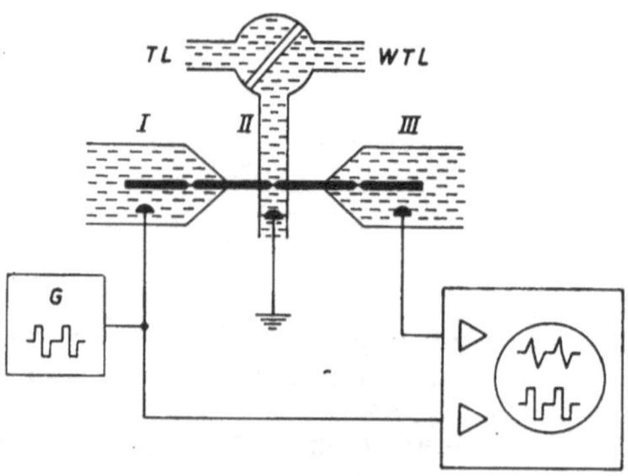

Abb. 1. Blockschaltbild der Versuchsanordnung.
G = Generator; rechts Zweistrahloszillograph mit Verstärkern; I, III Tröge mit Kokain-TASAKI-Lösung, Platinelektroden; II Schlauch mit bespültem Schnürring, Hahn, Platinelektrode; TL = TASAKI-Lösung; WTL = Wofapyrin-TASAKI-Lösung

Abbildung 1 zeigt die Versuchsanordnung. Der zu untersuchende Schnürring wurde in den Einschnitt eines Polyäthylenschlauches II gebracht. Ein Hahn im Zufluß des Schlauches gestattete das Umschalten auf Wofapyrinlösung. Die benachbarten Schnürringe lagen in je einem Trog aus Plexiglas. Der reizseitige Trog I war bis auf eine Öffnung für den Durchtritt der Nervenfaser allseitig abgeschirmt. Trog III auf der Ableitseite war gegen Trog I sorgfältig abgeschirmt. Als Elektroden tauchten Platinplättchen in die Tröge I und III sowie in den Zufluß und in den Abfluß II.

In den Trögen und zum Bespülen des Schnürrings verwendeten wir TASAKI-Lösung für Kaltblüter (vgl. v. MURALT [7]). Die Schnürringe in den Trögen I und III wurden mit Kokain-TASAKI-Lösung unerregbar gemacht. Zum Bespülen des Schnürrings II wurde der TASAKI-Lösung Wofapyrin zugesetzt, so daß eine Wofapyrinlösung der Konzentration 10^{-4} Mol/l entstand. Dies entspricht etwa der Konzentration im Blut bei der üblichen Dosierung. Alle Lösungen wurden mit demineralisiertem Wasser angesetzt.

Von einem Impulsgenerator, den VOLKMER [8] konstruiert und beschrieben hat, wurden im Abstand von etwa 2 sec Impulse als Reizspannung U zur Reizelektrode geführt. Der RANVIERsche Schnürring wurde mit kathodisch beginnenden, nullsymmetrischen Wechselimpulsen von 0,12 msec Dauer gereizt. Es wurden einzelne Wechselimpulse oder Serien von Wechselimpulsen mit einem Impulsabstand von 0,5 msec, also mit einer Reizfrequenz von 2 kHz dargeboten. Abbildung 7 zeigt eine Wechselimpulsgruppe, die in der dargestellten Polarität mit einem negativen Impuls an der geerdeten Elektrode II beginnt.

In der schematischen Darstellung der Nervenfaser in Abbildung 8 wurde nur der Kernwiderstand R_K (Längswiderstand) der Faser im reizseitigen Luftspalt gezeichnet, weil an ihm der größte Teil der Reizspannung $U(t)$ abfällt; der Membranwiderstand R_M (Querwiderstand) des untersuchten Schnürrings im Schlauch II ist viel kleiner als R_K. Das abgeleitete Potential $u(t)$ zeigt eine passive Verformung der Spannung $U(t)$ am komplexen Membranwiderstand $R_M||C_M$.

Ergebnisse

Zunächst wurden die Schwellenspannungen einzelner Wechselimpulse ermittelt. Im Abstand von einer Minute wurde die Spannung eingestellt, bei der auf einen Impuls hin ein Aktionspotential zustande kam. In der einen Meßreihe wurde der Schnürring mit normaler Tasaki-Lösung (TL), in der anderen mit Wofapyrin–Tasaki-Lösung (WTL) bespült, wobei zuvor 5 min lang unter Bespülung mit TL gemessen wurde. Die Mittelwerte aus mehreren Serien sind in Abbildung 2 zusammengestellt worden. Die zum Auslösen eines Aktionspotentials nötige Spannung stieg an, bei Bespülung mit WTL steiler als bei Bespülung mit TL.

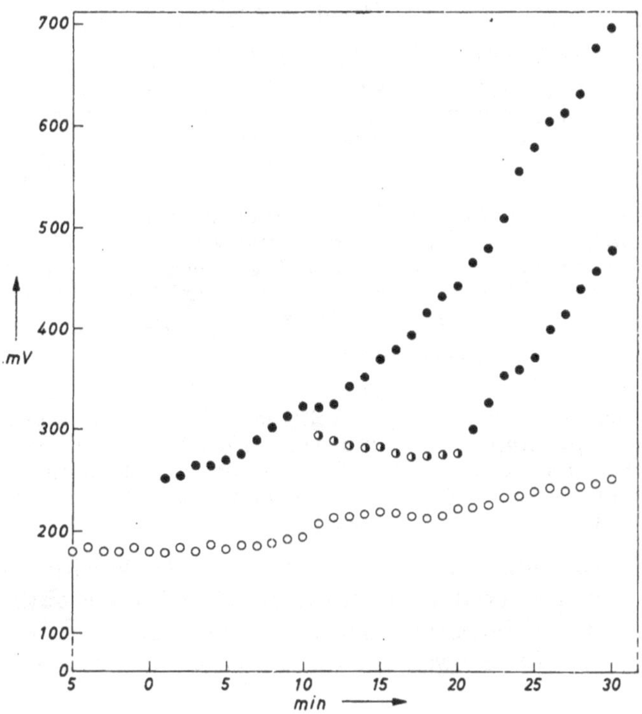

Abb. 2. Schwellenspannung U einzelner Wechselimpulse, Anstieg von U im Lauf der Zeit und unter Wofapyrinwirkung.

Abszissen: Zeit; Ordinaten: Reizspannung U. Weiße Punkte: Schnürring mit TL bespült; schwarze Punkte: mit WTL bespült; schwarz-weiße Punkte: nach vorheriger WTL-Bespülung mit TL bespült

Das Ansteigen der Schwellenspannung bei Bespülung mit TL ist eine Erscheinung, die man immer am isolierten, überlebenden Präparat findet; sie beruht wohl auf Absterbevorgängen im Zellstoffwechsel.

In einer weiteren Meßreihe wurde nach einer 10 min dauernden Bespülung mit WTL auf TL umgeschaltet: Die Schwellenspannungen nahmen jetzt ab, um wieder steil anzusteigen, nachdem nach weiteren 10 min wieder auf WTL umgeschaltet worden war.

Bereits in der 4. Minute der Bespülung mit WTL ist die Schwelle so angestiegen, daß ein signifikanter Unterschied zwischen den Schwellenspannungen bei Beispülung mit WTL und mit TL besteht; aus Tabelle 1 (a) geht dies hervor.

Tabelle 1

Statistischer Vergleich zwischen den Schwellenspannungen einzelner Wechselimpulse (a) und zwischen den relativen Schwellenamplituden von Gruppen aus 5 Wechselimpulsen (b) bei Bespülung mit TASAKI-Lösung und Wofapyrin-TASAKI-Lösung

Bespülung	n	$\bar{x}$	s_x	v_x	$s_{\bar{x}}$	d	s_d	t
a) normal Wofapyrin	99 68	187 265	41,8 130,8	22,4 49,3	4,2 15,8	78	14,1	5,5
b) normal Wofapyrin	25 18	86,5 82,6	2,94 2,48	3,39 3,03	0,59 0,58	3,9	0,83	4,7

Anmerkung: n = Zahl der Messungen; $\bar{x}$ = arithmetisches Mittel aus n Messungen; s_x = mittlere quadratische Abweichung der Einzelmessung; v_x = mittlere Abweichung in Prozent des arithmetischen Mittels, Variabilitätskoeffizient; $s_{\bar{x}}$ = mittlerer Fehler des Mittelwertes; d = Differenz zwischen den $\bar{x}$ der verglichenen Gruppen; s_d = mittlerer Fehler der Differenz; $t = d/s_d$.

$\bar{x}$, s_x, $s_{\bar{x}}$, d, s_d bei a in Millivolt, bei b in Prozent

Wegen des steilen Anstiegs der Reizschwelle unter Wofapyrinwirkung konnten die Messungen mit Impulsgruppen zur Untersuchung der Summation nicht mit aufeinander folgenden verschieden langen Impulsgruppen vorgenommen werden, die zu i–t-Kurven (Summationskurven) führten (SCHWARZ und VOLKMER [1]: Abb. 7), sondern es mußte anders vorgegangen werden:

Nach einer Schwellenmessung mit einem einzelnen Wechselimpuls wurde eine Impulsgruppe und danach wieder ein Einzelimpuls dargeboten. Abbildung 3 zeigt die Mittelwerte aus solchen Messungen, wobei Gruppen aus 2 Wechselimpulsen auf Schwellenspannung eingestellt worden waren. Die

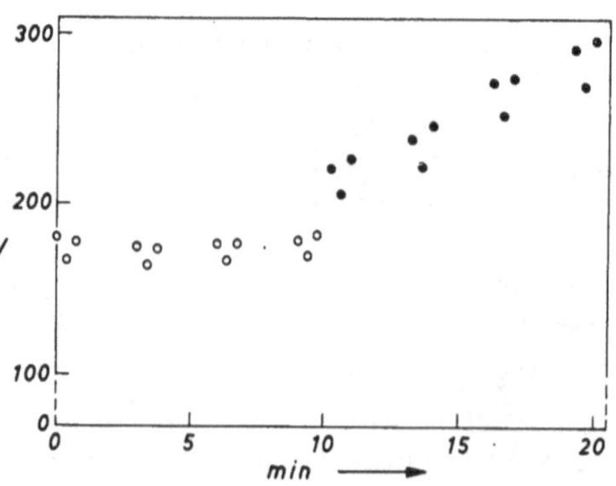

Abb. 3. Schwellenspannung von Impulspaaren und Einzelimpulsen am normalen Schnürring und unter Wofapyrinwirkung.

Abszissen: Zeit; Ordinaten: Reizspannung U. Weiße und schwarze Punkte: wie in Abbildung 2

Bespülung mit TL und WTL wurde in dieser Versuchsreihe an den gleichen Nervernfasern vorgenommen. Die Schwellenspannungen für die Impulsgruppen liegen niedriger als die für Einzelimpulse; dies ist eine Folge der Summation. Setzt man die Schwelle des Einzelimpulses = 100, dann liegt die Schwelle für die Gruppe aus 2 Wechselimpulsen unter normalen Bedingungen bei 93,9, unter Einwirkung von Wofapyrin bei 92,5; dies zeigt auch Abbildung 6.

In einer anderen Versuchsreihe wurde jeweils zwischen 2 Einzelreizen eine Gruppe von 5 Wechselimpulsen dargeboten, die Schwellenbestimmungen (1; 5; 1 Impulse) wurden alle 2 min vorgenommen.

Den Messungen unter Wofapyrineinwirkung gingen Schwellenbestimmungen unter Bespülung mit TL voraus. Meßreihen mit normaler Spülflüssigkeit und solche mit Wofapyrinbespülung liefen parallel.

Die Schwelle des Einzelimpulses wurde wieder = 100 gesetzt. Es zeigte sich, daß die Schwellen der Impulsgruppen unter Wofapyrineinwirkung niedriger liegen als unter normalen Bedingungen, wie aus Abbildung 4 hervorgeht. Die arithmetischen Mittel der Schwellenwerte sind in Abbildung 6 eingetragen worden. Der Unterschied zwischen beiden Schwellenwerten ist signifikant; dies ergibt sich aus Tabelle 1 (b).

Abb. 4. Relative Schwellenspannung von Gruppen aus 5 Impulsen am normalen Schnürring und unter Wofapyrinwirkung.

Abszissen: Zeit; Ordinaten: relative Reizspannung (für 1 Wechselimpuls = 100). Weiße und schwarze Punkte: wie in Abbildung 2

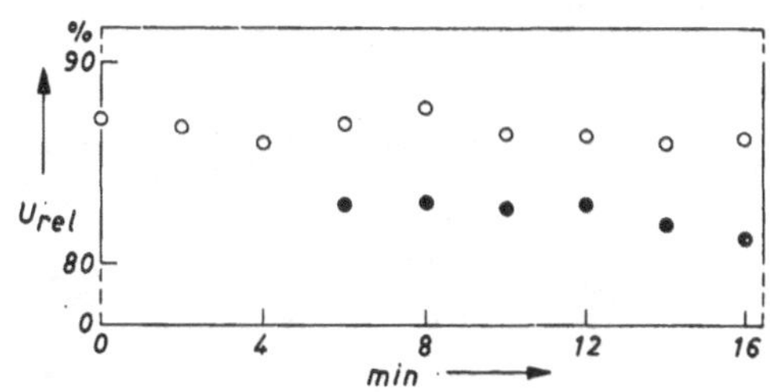

Die „Hauptnutzzeit" („HNZ") in dem auf den rhythmischen Reiz übertragenen Sinn (SCHWARZ [9]) ist infolge der Summation lokaler Potentiale länger als ein Impuls. An normal bespülten, frisch präparierten Schnürringen fanden wir bei Reizung mit einer Frequenz von 2 kHz als Mittelwert eine „HNZ" von 2,62 msec, nach 6 Wechselimpulsen kam es also zum Ausklinken eines Aktionspotentials[1]. Im Laufe der Zeit wurde die „HNZ" kürzer; dies beruht wahrscheinlich auf Absterbevorgängen im isolierten Präparat. Schürringe, die unter Wofapyrinwirkung standen, zeigten eine längere „HNZ", die im Laufe einer Meßreihe oft zunahm, mitunter auch längere Zeit konstant blieb. Abbildung 5 zeigt 2 Meßreihen, die von 2 Präparaten eines Tieres stammen.

Das Aktionspotential klinkt nach dem Ablauf von mehr oder weniger Wechselimpulsen aus, je nachdem die Reizspannung der Impulse kleiner oder größer ist. Abbildung 6 zeigt den Zusammenhang zwischen Impulszahl und

[1] Zur Berechnung der „HNZ" in msec s. SCHWARZ [10].

Schwellenspannung. Die Schwellenspannung eines einzigen Wechselimpulses wurde = 100 gesetzt. Bei Bespülung mit WTL sinkt die Summationskurve steiler ab als bei Bespülung mit TL.

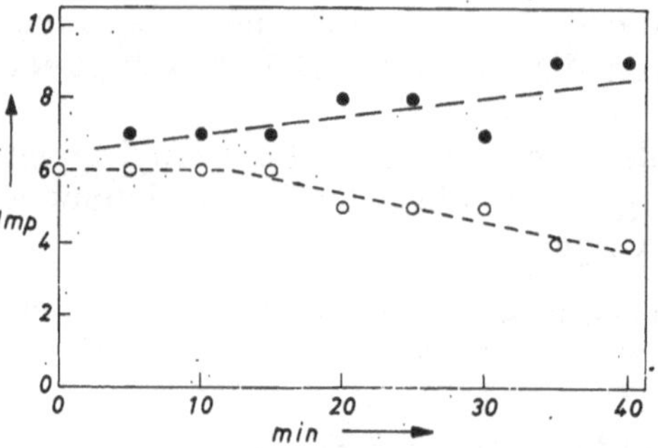

Abb. 5. „Hauptnutzzeiten". Versuch vom 14. 1. 66; 2 Präparate von *einem* Tier.
Abszissen: Zeit; Ordinaten: Zahl der Wechselimpulse pro „HNZ"-Gruppe. Weiße und schwarze Punkte: wie in Abbildung 2

Die Schwellenspannung für Impulsgruppen von der Länge der „HNZ" konnte unter Wofapyrinwirkung nur annähernd bestimmt werden, weil sie sich schnell ändert; der steile Anstieg in Abbildung 2 und 3 zeigt dies.

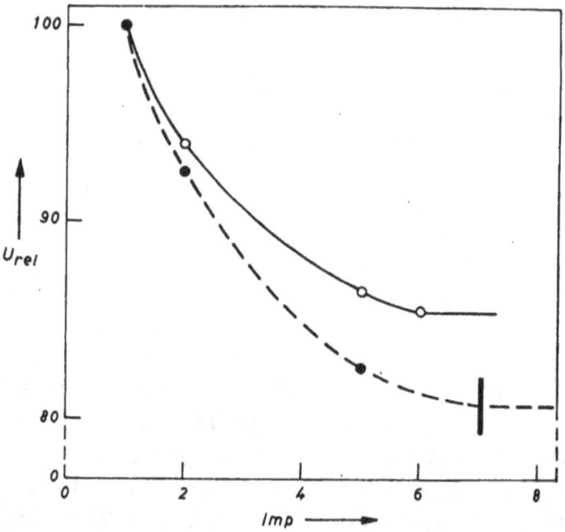

Abb. 6. Die „Nutzzeit" in Abhängigkeit von der Reizspannung.
Abszissen: Impulszahl; Ordinaten: relative Reizspannung. Weiße und schwarze Punkte: wie in Abbildung 2. Weißer Punkt bei 6 und schwarzer Balken bei 7 Impulsen: „Hauptnutzzeiten"

Diskussion

Wie Abbildung 7 zeigt, werden bei der Reizung der Nervenfaser mit Wechselimpulsen $U(t)$ am Schnürring verformte Potentiale $u(t)$ abgeleitet. Die Reizimpulse werden durch den komplexen Widerstand des Schnürrings $R_M || C_M$ nur passiv verformt, wenn die Reizspannung wesentlich unter der

Schwellenspannung liegt; Abbildung 7a zeigt $u_p(t)$. Bei höherer Reizspannung zeigen sich kurz vor dem Erreichen der Schwellenspannung stärkere Verformungen der abgeleiteten Potentiale $u(t)$, die durch das Hinzukommen eines aktiven Anteils $u_a(t)$ gedeutet werden können (Abb. 7b). Der aktive Anteil läßt sich durch Subtraktion der passiven Potentialschwankung vom abgeleiteten Potential ermitteln (Abb. 7c).

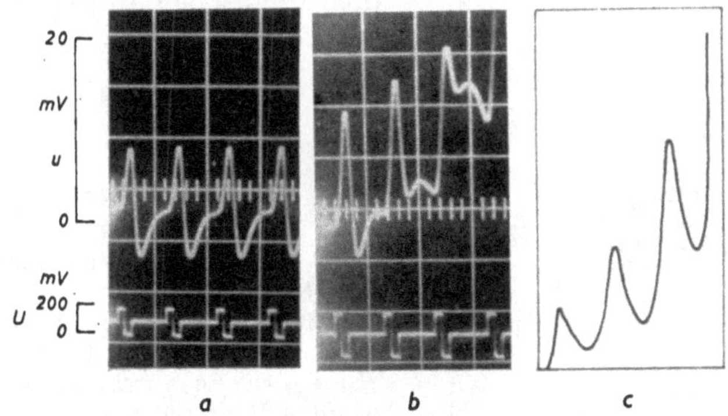

Abb. 7. Impulsgruppen einer Frequenz von 2 kHz.
Unten Reizspannung, oben abgeleitete Potentiale. a = weit unterschwellig; b = schwellig, mit dem 4. Impuls klinkt ein Aktionspotential aus; c = aktiver Potentialanteil im Oszillogramm b; Differenzkurve $u_a(t) = u(t) - 1{,}84 \cdot u_p(t)$

Das passiv verformte Potential weit unterschwelliger Impulse mußte in Richtung der Ordinate vergrößert werden, weil ja die stärkeren Wechselimpulse, die höhere lokale Potentialschwankungen bis zum Ausklinken eines Aktionspotentials hervorrufen, auch eine höhere passive Potentialänderung verursachen. Es ergab sich

$$u_a(t) = u(t) - 1{,}84 \cdot u_p(t)\,.$$

In den Mitteilungen von Schwarz und Volkmer [1] und Schwarz [2] ist das Verfahren ausführlich beschrieben worden.

Die Differenzkurve $u_a(t)$ (Abb. 7c) zeigt, daß die aktiven Potentialanteile nach jedem weiteren Wechselimpuls größer werden und daß eine immer größere Restspannung in Richtung einer Depolarisation bleibt. Ein Potential, das bei einer Depolarisation von etwa 20 mV beginnt, kann sich schließlich zu einem Aktionspotential entwickeln.

Abbildung 8 zeigt ein Äquivalentschaltbild der Nervenfaser. Darin ist als Ursache für den aktiven Potentialanteil ein „Generator" eingezeichnet worden, der durch Änderung der elektrischen Parameter am Schnürring (vielleicht der Ionenfluxe) eine „lokale Erregung" hervorbringt.

Bei Reizung mit Wechselimpulsen findet durch den anodischen Impuls, der jedem kathodischen folgt, eine Repolarisation statt, gegen die die Summation durch zunehmende Depolarisation mit Hilfe des aktiven Anteils erfolgen muß. Unter Wofapyrineinwirkung ist die Summation wirksamer als bei Bespülung mit Tasaki-Lösung; für eine Impulsgruppe genügt, prozentual gerechnet, eine geringere Reizspannung, um ein Aktionspotential ausklinken zu lassen (Abb. 4). Man könnte daraus schließen, daß der aktive Anteil am lokalen Potential, die „lokale Erregung", durch Wofapyrin vergrößert wird. Die Notwendigkeit des größeren aktiven Potentialanteils könnte ihren Grund darin haben, daß Wofapyrin die Repolarisation durch den anodischen Impuls begünstigt.

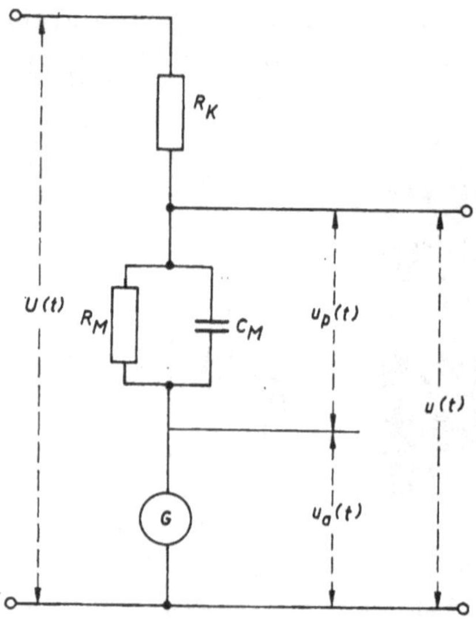

Abb. 8. Äquivalentschaltbild der Nervenfaser.

R_K = Kern(Längs-)Widerstand des Nerven im reizseitigen Luftspalt; $U(t)$ = Reizspannung; $u(t)$ = abgeleitetes Potential; $u_p(t)$ = durch den Komplexwiderstand $R_M||C_M$ der Schnürringmembran passiv verformtes Potential bei weit unterschwelliger Reizung; $u_a(t)$ = vom „Generator" G aktiv erzeugter Potentialanteil

Unter Wofapyrin erstreckt sich die Summation über eine größere Zahl von Impulsen als bei Bespülung mit TASAKI-Lösung, wenn mit Impulsgruppen von der Dauer der „Hauptnutzzeit" gereizt wird. Der aktive Anteil wird offenbar in einer größeren Zahl von Schritten zu größerer Höhe aufgebaut, um ein Aktionspotential ausklinken zu lassen (Abb. 5 und 6).

Wie aus Abbildung 4 hervorgeht, genügt zwar unter Wofapyrinwirkung für eine Impulsgruppe prozentual eine geringere Reizspannung zum Ausklinken eines Aktionspotentials, die absolute Reizspannung ist aber größer als beim normal bespülten Schnürring, wie Abbildung 3 zeigt. Letzteres gilt auch für einzelne Wechselimpulse (Abb. 2). Dies könnte man damit deuten, daß Wofapyrin die erregbare Membran verfestigt, so daß zur Depolarisation höhere Reizspannungen benötigt werden.

Aus Abbildung 6 geht hervor, daß zu einer „HNZ"-Impulsgruppe beim normalen Schnürring 6, unter Wofapyrinwirkung 7 Impulse gehören. Ohne Wofapyrinwirkung würden 7 Impulse pro „HNZ" einer Reizfrequenz von 2,2 kHz, die Amplitude 80,7 der „HNZ"-Impulsgruppe (Abb. 6, gestrichelte Waagerechte) einer Frequenz von 2,8 kHz entsprechen. Der unter Wofapyrinwirkung stehende Schnürring verhält sich so, als ob er nicht von 2 kHz, sondern von etwa 2,5 kHz gereizt würde, er ist also „langsamer" geworden.

Literatur

[1] SCHWARZ, F. u. H. VOLKMER: Acta biol. med. german. **15**, 283 (1965).
[2] SCHWARZ, F.: Acta biol. med. german. **16**, 625 (1965).
[3] STÄMPFLI, R.: Ergebn. Physiol., biol. Chem. exp. Pharmakol. **47**, 70 (1952).

[4] Stämpfli, R.: J. de Physiol. **48**, 710 (1956).
[5] Kilb, H. u. R. Stämpfli: Helv. physiol. pharmacol. Acta **14**, 251 (1956).
[6] Rudolph, G. u. R. Stämpfli: Pflügers Arch. ges. Physiol. Menschen Tiere **267**, 524 (1958).
[7] Muralt, A. v.: Neue Ergebnisse der Nervenphysiologie. Springer-Verlag, Berlin-Göttingen-Heidelberg 1958.
[8] Volkmer, H.: Acta biol. med. german. **3**, 193 (1959).
[9] Schwarz, F.: Pflügers Arch. ges. Physiol. Menschen Tiere **261**, 361 (1955).
[10] Schwarz, F.: Acta biol. med. german. **3**, 200 (1959).

Anschrift der Verfasser: Prof. Dr. Friedrich Schwarz und Dr. Rosemarie Thiele, Physiologisches Institut der Universität, 69 Jena, Teichgraben 8.

Summary

F. Schwarz and R. Thiele: Effect of a combination of amidopyrine and phenylbutazone (wofapyrine) on Ranvier's nodes of nerves of cold-blooded animals

Ranvier's nodes of motoric nerve fibres from the Nervus ischiadicus of frogs are stimulated with alternating electric square pulses. Subthreshold local potentials and action potentials are tapped. The stimulating pulses cause local changes to the potential, which are summing-up. The node is wetted with wofapyrin.
The stimulus threshold of the individual alternating pulses rises steeply under the effect of wofapyrin; this effect can be partially balanced by rinsing. The threshold voltages for pulse groups of 2 kc are lower than for single pulses—as a consequence of summation. The summation is higher under the effect of wofapyrin than with the normally wetted node.

Резюме

Ф. Шварц и Р. Тиле: О действии вофапирина на сужения Ранвье нервных волокон хладнокровных животных

Сужения Ранвье двигательных нервных волокон из седалищного нерва лягушки стимулировали прямоугольными электрическими переменными импульсами. Отводились подпороговые локальные потенциалы и потенциалы действия. Каждый стимул вызывает локальные изменения потенциала, которые суммируются. Сужения нервных волокон омываются раствором вофапирина.
Порог раздражения отдельного переменного импульса резко повышается под действием вофапирина. Этот эффект можно частично устранить промывкой. Пороговые напряжения для группы импульсов, имеющей частоту в 2 кгц, ниже, чем для отдельных импульсов, вследствие суммации. Суммация оказалась сильнее под действием вофапирина, чем у нормально омываемого сужения.
При пороговом напряжении регистрируется количество импульсов, необходимое для создания потенциала действия. Количество импульсов выше под действием вофапирина.
Вофапирин снижает раздражительность сужений нервных волокон, возможно благодаря упрочению раздражительных мембран.

Acta biol. med. german., Band 17, Seite 96—98 (1966)

Aus dem Physiologischen Institut der Friedrich-Schiller-Universität Jena
(Direktor: Prof. Dr. F. Schwarz)

Membranpotentiale alternder Muskelzellen

F. Schwarz und I. Wichan

(Eingegangen am 4. 3. 1966)

Zusammenfassung

Von Skelettmuskelfasern der Weißen Ratte wurden mit Glasmikroelektroden Membranpotentiale abgeleitet. Bei jungen Tieren ist das Membranpotential niedrig. Es steigt mit zunehmendem Alter an und hat mit etwa 80 Tagen seine volle Höhe erreicht. Im Greisenalter, mit 1,5—2 Jahren, nimmt das Membranpotential wieder ab.
Der Variabilitätskoeffizient ist bei 35 Tage alten Tieren besonders groß. Dies deutet auf einen unterschiedlichen Reifegrad der Muskelfasern hin.

Es ist bekannt, daß sich der Stoffwechsel und die stoffliche Zusammensetzung der Muskelzelle mit dem Alter ändern (Bürger [1]). So nimmt die Aufspaltbarkeit des Aktomyosins durch Adenosintriphosphat mit zunehmendem Alter beträchtlich ab (Wielepp [2]), der Wassergehalt und der Gehalt von Kalium und Kalzium werden geringer (Lohmann und Mitarbeiter [3]). Da Membranpotential und Stoffwechsel der Zelle eng miteinander verknüpft sind, liegt es nahe, auch an Änderungen des Membranpotentials mit zunehmendem Alter zu denken. Im Schrifttum fanden wir darüber nichts. In der vorliegenden Untersuchung ist das Ruhepotential der Skelettmuskelfasern von Ratten verschiedenen Alters gemessen worden.

Methodik

Weiße Ratten im Alter zwischen 22 Tagen und etwa 2 Jahren, männliche und weibliche Tiere, wurden mit 25%iger Äthylurethanlösung (0,4 bis 2,2 ml i.p. dem Alter entsprechend) narkotisiert. Die Oberschenkelmuskulatur wurde freigelegt, die Faszie entfernt. Die Mikroelektroden wurden in Muskelfasern der Adduktoren eingestochen. 52 Tiere wurden verwendet, an jedem Tier wurden etwa 80 Messungen vorgenommen.
Die Ruhepotentiale wurden mit Glasmikroelektroden abgeleitet, wie sie Volkmer [4] beschrieben hat. Eine Makroelektrode war als Bezugselektrode in der Nähe der Meßstelle auf die Muskulatur aufgesetzt worden. Die Potentiale wurden an einem *p*H-Meßverstärker (MV 11, Clamann und Grahnert, Dresden) abgelesen (Schwarz und Mitarbeiter [5], Kunze [6]).

Ergebnisse

In Abbildung 1 sind die Befunde dargestellt worden. Bei sehr jungen, 22 Tage alten Tieren ist das Membranpotential noch niedrig (67,9 mV). Es steigt mit zunehmendem Alter steil an und hat im Alter von etwa 80 Tagen, wenn die Tiere geschlechtsreif sind, seine volle Höhe (90,3 mV) erreicht. Im Greisenalter, mit 1,5—2 Jahren, nimmt das Ruhepotential wieder ab (auf 86,2 mV). Die Mittelwerte für die verschiedenen Altersstufen unterscheiden sich signifikant voneinander.

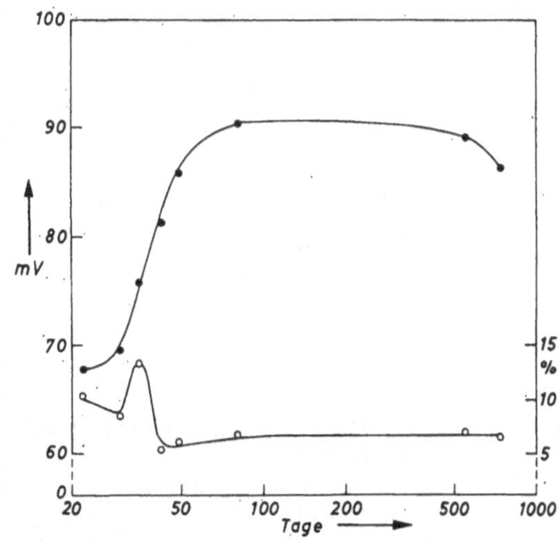

Abb. 1. Membranpotentiale von der Schenkelmuskulatur der Ratte (schwarze Punkte) und Variabilitätskoeffizienten (weiße Punkte) in Abhängigkeit vom Alter.

Abszissen (logarithmisch): Zeit (Tage); Ordinaten: Membranpotential (Millivolt); Variabilitätskoeffizient (%)

Bei der statistischen Bearbeitung der Meßergebnisse fiel auf, daß die mittlere quadratische Abweichung der Einzelmessung (standard deviation) bei den 35 Tage alten Tieren besonders groß ist. Der Variabilitätskoeffizient v_x (mittlere Abweichung in Prozent des arithmetischen Mittels) zeigt dies noch besser, er ist deshalb für die verschiedenen Altersstufen in Abbildung 1 eingetragen worden.

Die große Streuungsbreite bei 35 Tage alten Tieren könnte ihre Ursache einmal darin haben, daß diese Tiere ein unterschiedliches biologisches Alter haben, daß etwa die einen zu den 22 bis 30 Tage alten, die anderen zu den 42 bis 49 Tage alten Tieren gerechnet werden müßten. Zum anderen könnten aber auch bei jedem einzelnen Individuum die Muskelfasern einen unterschiedlichen Reifegrad haben.

Die arithmetischen Mittel der Membranpotentiale und die Variabilitätskoeffizienten für jedes einzelne Tier wurden aus den etwa 80 Messungen je Tier ermittelt und in Tabelle 1 zusammengestellt, wobei die Mittelwerte nach der Größe geordnet wurden. Man sieht, daß bei jedem 35 Tage alten Tier v_x größer ist als bei jedem 49 Tage alten. Daraus kann man schließen, daß die Muskelfasern der 35 Tage alten Tiere weniger einheitliche Membranpotentiale und damit wohl auch unterschiedlichere Membranstrukturen haben als die Muskelfasern der älteren Tiere, der Reifegrad der Muskelzellen scheint eine größere Schwankungsbreite zu haben.

Tabelle 1
Individuelle Mittelwerte $\bar{x}$ der Membranpotentiale und Variabilitätskoeffizienten v_x von Ratten verschiedener Altersstufen

22 Tage alt		35 Tage alt		49 Tage alt	
$\bar{x}$	v_x	$\bar{x}$	v_x	$\bar{x}$	v_x
63,8	11,30	71,3	15,45	83,4	6,23
65,8	11,87	75,2	10,70	83,8	7,27
66,2	6,83	75,6	13,45	84,9	5,89
68,1	9,55	76,0	12,48	85,4	5,40
68,4	9,68	77,2	13,65	86,1	4,22
69,2	11,15	77,5	13,04	86,5	6,06
71,7	12,90	80,6	12,10	87,3	6,94

Auch die jüngsten untersuchten Tiere im Alter von 22 Tagen weisen größere Variabilitätskoeffizienten auf (Tab. 1) als die 30, 42, 49 Tage alten Tiere (Abb. 1). Die Ursache dafür scheint auch hier der unterschiedliche Reifegrad der Muskelfasern zu sein.

Literatur

[1] BÜRGER, M.: Altern und Krankheit als Problem der Biomorphose. 4. Aufl. VEB Georg Thieme, Leipzig 1960.
[2] WIELEPP, G.: Chemische Untersuchungen über Alternsveränderungen am menschlichen Skelettmuskel. Inaug. Diss. Leipzig 1955.
[3] LOHMANN, D., G. WIELEPP u. J. NÖCKER: Z. Alternsforsch. **10**, 233 (1957).
[4] VOLKMER, H.: Acta biol. med. german. **15**, 348 (1965).
[5] SCHWARZ, F., M. BIEDERMANN u. H. VOLKMER: Acta biol. med. german. **15**, 608 (1965).
[6] KUNZE, M.: Acta biol. med. german. **17**, 48 (1966).

Anschrift der Verfasser: Prof. Dr. FRIEDRICH SCHWARZ und IRENE WICHAN, Physiologisches Institut der Universität, 69 Jena, Teichgraben 8.

Summary

F. SCHWARZ and I. WICHAN: Membrane potentials of aging muscle cells

Membrane potentials are tapped from skeletal muscle fibers of albino rats using glass microelectrodes. Membrane potentials are low with young animals, increasing at higher age, reaching its full level at about 80 days, and falling again in senility, between 1,5 and 2 years.
The coefficient of variability is particularly high with 35-days-old animals, indicating a different maturity of the muscle fibers.

Резюме

Ф. Шварц и И. Вихан: Мембранные потенциалы стареющих мышечных клеток

Отводились мембранные потенциалы из волокон скелетных мышц белой крысы при помощи стеклянных микроэлектродов. Уровень мембранного потенциала оказался низким у детенышей, повышался с повышением возраста животных и достиг полной высоты к 80-му дню. В возрасте 1,5 до 2 лет мембранный потенциал опять падал.
Коэффициент изменчивости особенно велик в возрасте 35 дней, что указывает на различную зрелость мышечных волокон.

Acta biol. med. german. Band 17, Seite 99—115 (1966)

Aus dem Institut für Pathologische Physiologie der Friedrich-Schiller-Universität Jena
(Direktor: Prof. Dr. E. GOETZE)

Elektronenmikroskopische Untersuchungen über „helle und dunkle" Leberzellen in der regenerierenden Rattenleber

H. FRANKE und E. GOETZE

(Eingegangen am 8. 3. 1966)

Zusammenfassung

Es wurde die Substruktur der Rattenleber 4—96 Std. nach Teilhepatektomie untersucht. Der Verlauf der Regeneration ist mit charakteristischen feinstrukturellen Veränderungen verbunden. Anfangs kommt es zu einer gleichmäßigen intensiven Leberverfettung und am endoplasmatischen Retikulum zu Vakuolisierungserscheinungen, die von einer allgemeinen Reduktion des Ergastoplasmas begleitet sind bei gleichzeitigem Auftreten zahlreicher freier Ribosomen.

Besonders fällt auf, daß neben hellen Leberzellen in vermehrter Anzahl dunkle Leberzellen vorhanden sind, die sich durch eine teils sehr hohe Elektronendichte und ebenfalls durch einen hohen Gehalt an freien Ribosomen auszeichnen.

Während sich in den hellen Leberzellen die feinstrukturellen Veränderungen relativ schnell normalisieren, bleibt die charakteristische Substruktur in den dunklen Leberzellen erhalten. Ihre Anzahl geht nur langsam zurück, hierbei ergaben sich keine Hinweise einer Degeneration.

Die elektronenoptischen Befunde an der regenerierenden Leber und der von Ratteneugeborenen sprechen dafür, daß die dunklen Leberzellen in engem Zusammenhang mit einer erhöhten Zellproliferation stehen und daß sie sich im Verlaufe der Regeneration wieder in helle Zellen umbilden können.

1892 beschrieb COHN [1] in der Leber von Katze, Maus, Hund und Kaninchen zwei verschiedene Zellarten: Zellen mit hellem Zelleib und solche mit homogenem, tiefdunkel aussehendem Zelleib. In den nachfolgenden Jahren ist über das Vorkommen von dunklen Leberzellen (DLZ) und hellen Leberzellen (HLZ) in der Leber verschiedener Säuger wiederholt in experimentellen licht- und elektronenmikroskopischen Untersuchungen berichtet worden, wobei die DLZ in vermehrter Zahl besonders nach Einwirkung schädigender Faktoren auf die Leber beobachtet wurden.

So unter anderem nach absoluter oder teilweiser Hungereinwirkung [2]—[6] oder nach temporärer Durchblutungssperre bzw. O_2-Mangel [7]—[9]. Auch nach Applikation verschiedener, meist leberschädigender Verbindungen wurden DLZ gesehen, wie nach Äthionineinwirkung [6], [10], [11] oder nach Einwirkung von CCl_4, Thioazetamid, α-Naphthylisothiozyanat oder Tanninsäure [12]—[14]. STEINER et al. [15] berichtet auch über das Vorkommen von DLZ nach Gallengangligatur. ATERMAN [16] sowie FISCHLER und ROECKL [3] stellten nach hormoneller Beeinflussung der Leber DLZ in vermehrtem Umfang fest. BRAUN [17] berichtet von einer steigenden Zellverdichtung in der Rattenleber nach Einwirkung höherer Röntgendosen. Diesen Befund konnten wir [18] für die röntgenbestrahlte Leber gravider Ratten bestätigen, außerdem beobachteten wir DLZ in der Leber gravider Ratten nach experimenteller Durstexsikkose [19].

Bei der Mehrzahl der elektronenmikroskopischen Untersuchungen ist der feinstrukturelle Befund über die DLZ etwa gleichlautend. Es handelt sich im typischen Fall um Zellen mit einem sehr elektronendichten Grundzytoplasma, in dem in der Regel die Zellorganellen und Membranen des endoplasmatischen Retikulums (ER) sehr eng beisammen liegen, so daß sie pro Flächeneinheit zahlreicher erscheinen als in den normalen Leberzellen. Auf Grund etwa gleicher Dichteverhältnisse des Grundzytoplasmas zeichnen sich die Zellorganellen und Membranen des ER sowie auch die Zellkerne in den DLZ nur undeutlich von ihrer Umgebung ab. Charakteristisch ist weiterhin der häufig sehr hohe Gehalt an freien Ribosomen, der den DLZ bzw. dem Grundzytoplasma einen granulären Aspekt verleiht.

Diesen typischen Dunkelzellen stehen die HLZ gegenüber, sowie weniger dunkle Zellen, die eine Art Übergangsform zwischen beiden Extremen darstellen. Derartige Übergangsformen können auch in der normalen Säugerleber beobachtet werden. Es handelt sich hierbei um Zellen, deren dunkleres Aussehen vor allem durch stärkere Glykogeneinlagerungen, erhöhten Ferritingehalt oder auch durch einen besonderen Reichtum an Zellorganellen bedingt ist.

Die auch heute noch sehr unterschiedliche Bewertung der typischen DLZ sowie unsere im Zusammenhang mit anderen Untersuchungen wiederholten Beobachtungen eines vermehrten Vorkommens von DLZ in der normalen Leber wenige Tage alter, gesunder Ratten [20], zu einem Zeitpunkt also, wo noch eine starke Zellvermehrung stattfindet, waren Anlaß zu einer genauen Untersuchung der Zellverhältnisse in der regenerierenden Rattenleber.

Material und Methodik

Erwachsenen männlichen Ratten eines Wistar-Inzuchtstammes, Gewicht etwa 200 g, wurde nach Laparotomie in Äthernarkose der große Lebermittellappen nach vorheriger Unterbindung reseziert. Alle Tiere bekamen eine Standardkost, wobei darauf geachtet wurde, daß sie nach der Operation fraßen, um Hungereffekte auszuschließen. Die Fixierung der Leberproben erfolgte 4, 24, 48 und 96 Std. nach Teilhepatektomie in OsO_4 nach der Methode von CAULFIELD [21]. In allen Fällen wurden unter gleichen technischen Bedingungen Leberproben von Kontrolltieren untersucht. Die Einbettung des Materials erfolgte in Vestopal W [22], die Anfertigung der Dünnschnitte mit einem PORTER-BLUM-Mikrotom, ihre Durchsicht im Elmi D2 des VEB Zeiss bei einer Originalvergrößerung bis zu 12790fach. Ein Teil der Schnitte wurde nach der Methode von KARNOVSKY [23] kontrastiert.[1]

Um ein genaueres Bild über die Menge und Verteilung der Lipideinschlüsse in den Leberazini zu bekommen, wurden etwa 6 mm^3 große Leberwürfel 5 Tage lang osmiert und danach Schnitte für lichtmikroskopische Untersuchungen angefertigt.

[1] Herrn Prof. Dr. F. BOLCK, Pathologisches Institut, sowie Herrn Dr. W. SCHWARTZE von der Forschungsstelle für Übermikroskopie der Universität Jena sind wir auch dieses Mal wieder zu Dank verpflichtet für die Mitbenutzung des Elektronenmikroskops, ebenso Herrn Prof. Dr. G. GEYER, Anatomisches Institut der Universität Jena, für die Mitbenutzung des Ultramikrotoms

Ergebnisse

Die Übersichtsabbildung 1 zeigt einen Befund *4 Std. nach Teilhepatektomie.* Im Vergleich zum Normalbild fällt als erstes der außerordentlich hohe Lipidgehalt in den Leberzellen auf. Wie aus gleichzeitig durchgeführten lichtmikroskopischen Untersuchungen an osmierten Leberschnitten hervorgeht, erstreckt sich diese abnorme Verfettung gleichmäßig über den gesamten Bereich der Leberazini. Da viele dieser tropfenförmigen Lipideinschlüsse teilweise einen größeren Durchmesser als die Leberzellkerne besitzen, sind sie lichtmikroskopisch bereits bei mittlerer Vergrößerung gut erkennbar.

Als weiterer hervorstechender Befund im Vergleich zu den Kontrollen muß die besondere Elektrodendichte einzelner Leberzellen oder ganzer Zellgruppen genannt werden. Diese unterschiedliche Dichte zwischen HLZ und DLZ ist auf Abbildung 1 erkennbar, wobei es sich hier um Übergangsstadien handelt, während Abbildung 2 ein Beispiel für eine typische DLZ ist. Die Verteilung der DLZ zwischen den HLZ läßt keine Regelmäßigkeit erkennen. Sie treten als Einzelzellen auf oder bilden inselartig Zellnester im Verband der immer in der Überzahl vorhandenen HLZ, gelegentlich sind sie auch randständig um die Sinusoide angeordnet. Bei den typischen Dunkelzellen, wie in Abbildung 2, ist die Elektronendichte so hoch, daß die Innenstrukturen der Zelle nicht mehr oder nur noch schwach erkennbar sind. Um die Feinstrukturen einigermaßen sicher beschreiben zu können, müssen aus diesem Grund vor allem Leberzellen mit einem nicht zu hohen Verdichtungsgrad herangezogen werden.

Schon auf den Übersichtsabbildungen (Abb. 1 und 2) ist erkennbar, deutlich zeigen es jedoch stärkere Vergrößerungen (Abb. 3), daß 4 Std. nach Teilhepatektomie sowohl in den DLZ als auch in den HLZ die Feinstrukturen zum Teil vom Normalbefund abweichen. Die Bereiche des Ergastoplasmas weisen eine vakuolenartige Erweiterung der zwischen den Membranen gelegenen Raumanschnitte auf (Abb. 2 u. 3). Diese sind elektronenoptisch leer. Auch das übrige ER zeigt deutlich Vakuolisierungserscheinungen. Bemerkenswert ist außerdem die große Anzahl freier Ribosomen, die neben besonders elektronendichten glykogenhaltigen Partikeln relativ gleichmäßig in der zytoplasmatischen Grundsubstanz verteilt liegen (Abb. 3). Im Gegensatz zu den Lamellensystemen des ER und trotz hoher Verfettung des Leberparenchyms sind an den Mitochondrien in den HLZ keine Veränderungen nachweisbar. Gleiches gilt für die Lysosomen und Mikrobodies sowie auch für die Zellkerne. Ein Vergleich der HLZ mit den DLZ ergibt deutlich, daß in den letzteren die Vakuolisierungserscheinungen weniger stark sind (Abb. 2 und 3). Infolgedessen und bedingt durch die dunkle zytoplasmatische Grundsubstanz machen die DLZ einen sehr kompakten Eindruck. Ähnlich wie in den HLZ sind ungewöhnlich viel freie Ribosomen neben Glykogenpartikeln vorhanden, wie sie in dieser Anzahl in der normalen Leberzelle nicht gefunden wurden. Bezüglich des Lipidgehaltes besteht zwischen beiden Zelltypen kein wesentlicher Unterschied, HLZ sowie DLZ zeigen in etwa gleichem Umfang Verfettungserscheinungen.

24 Std. nach Teilhepatektomie hat sich das Gesamtbild in Bezug auf das Vorkommen DLZ nicht wesentlich geändert. Alle untersuchten Leberproben enthalten zu diesem Zeitpunkt in vermehrter Anzahl typische Dunkelzellen und daneben vor allem auch solche Zellen, die bezüglich ihrer Elektronendichte eine Mittelstellung zwischen HLZ und DLZ einnehmen (Abb. 4). Es ist jetzt jedoch zu folgenden auffälligen feinstrukturellen Veränderungen gekommen, die beide Zelltypen in gleichem Maße betreffen: Erstens ist der vorangegangene abnorm hohe Gehalt an Lipideinschlüssen überraschend schnell fast auf das Normalmaß zurückgegangen. Elektronenmikroskopisch sichtbare Fettpartikel sind nur noch selten sichtbar, von geringem Durchmesser und oft von hellen Vakuolen durchsetzt. Sie machen einen in Auflösung befindlichen Eindruck (Abb. 5). Zweitens haben sich die Vakuolisierungserscheinungen am ER allgemein zurückgebildet und drittens sind im Vergleich zum 4 Std.-Befund und zur Normalleber die Ergastoplasmabereiche nur noch spärlich vorhanden. Soweit Ergastoplasmabereiche sichtbar sind, zeigen sie jetzt keine Dilatation ihrer Raumanschnitte mehr (Abb. 5). Bei beiden Zelltypen ist dagegen die Anzahl der vielen freien Ribosomen unverändert hoch geblieben, das gilt besonders für die DLZ.

48 Std. nach Teilhepatektomie ist die Zahl der DLZ noch immer erhöht. Neben diesen und den HLZ fallen vor allem auch zahlreiche Zellen von mittlerer Elektronendichte auf. Im Vergleich zum 24-Std.-Befund sind in den HLZ und auch in denjenigen von mittlerer Elektronendichte jetzt wieder viel häufiger kleinere oder größere Ergastoplasmabereiche anzutreffen. Alle Doppellamellen sind dicht mit Ribosomen besetzt (Abb. 6), daneben kommen aber auch viele freie Ribosomen vor, die entweder zwischen den Ergastoplasmamembranen oder in größerer Ansammlung im Grundzytoplasma liegen.

96 Std. nach Teilhepatektomie sind ebenfalls DLZ vorhanden. Im Vergleich zu den vorangegangenen Befunden sind sie aber deutlich seltener geworden. In den Fällen, in denen durch relativ wenig Ergastoplasma und weiträumige Verteilung der Zellorganellen das Bild der HLZ zusätzlich verstärkt wird, ist der Übergang zu einer benachbart liegenden typischen Dunkelzelle besonders krass ausgeprägt (Abb. 7). Wie aus Abbildung 7 weiterhin hervorgeht, sind auch 96 Std. nach Teilhepatektomie bei den typischen DLZ die ihnen charakteristischen Strukturbesonderheiten in vollem Umfang erhalten geblieben. Dieser Befund ist für die Beurteilung der Frage, ob es sich im Falle der DLZ um Degenerationsstadien handelt, sehr wesentlich.

Diskussion

Überblickt man die feinstrukturelle Reaktion der nach Teilhepatektomie verbliebenen Leberzellen, so fallen in den untersuchten Zeiträumen von 4 bis zu 96 Std. einige charakteristische Veränderungen besonders auf. Hierzu gehören die sehr intensive akute Leberverfettung, die akuten Vakuolisierungserscheinungen am ER, temporärer Reichtum an freien Ribosomen sowie das vermehrte Vorkommen typischer DLZ neben weniger dunklen und HLZ. Während die erstgenannten Erscheinungen, Verfettung und Vakuolisierung, sich relativ schnell normalisieren, sind die DLZ mit ihrer charakteristischen Struktur in allen Untersuchungsphasen nachweisbar, besonders aber in den ersten 4—48 Std. nach Teilhepatektomie.

Die Substruktur der Leber nach Teilhepatektomie wurde bereits von Aterman [24] sowie von Davis [25] bei Ratten, außerdem von Trotter [26] bei Mäusen untersucht, wobei die untersuchten Zeiträume z. T. voneinander abweichen. Ein Vergleich dieser Befunde miteinander und mit unseren Ergebnissen ergibt Übereinstimmungen, aber auch Abweichungen. Während wir nur in mäßigem Umfang Vakuolisierungsprozesse am ER feststellten, war bei den Atermanschen und Davisschen Untersuchungen die Rattenleber in den ersten Stunden nach Teilhepatektomie hochgradig vakuolisiert. Hierbei war es Davis [25] sogar möglich, im Verlaufe der ersten 24 Std. der Regeneration zwischen zwei gesondert auftretenden Vakuolisierungsphasen zu unterscheiden. Trotter [26] dagegen erwähnt nichts über Vakuolisierungserscheinungen in der regenerierenden Mäuseleber.

In ähnlicher Weise differieren die Befunde über die akute Leberverfettung. Davis stellte beim 12-Std.'-Befund ein vermehrtes Vorkommen nur kleiner Lipidkörper fest, bei Aterman fehlen Angaben über besondere Abweichungen im Lipidgehalt gänzlich, während Trotter berichtet, daß in der Mäuseleber das Maximum der elektronenmikroskopisch sichtbaren Lipide erst zwischen 16 und 48 Std. eintritt. Im Gegensatz zu unseren Befunden handelt es sich dabei nur um relativ wenige und sehr kleine Lipideinschlüsse. Es fällt weiterhin auf, daß in keiner der genannten Arbeiten das Vorkommen von DLZ erwähnt wird.

Wie bereits betont, sind die Meinungen über die DLZ z. T. sehr unterschiedlich. Daß es sich bei ihnen nicht um Fixierungsartefakte handelt, darüber ist man sich inzwischen wohl allgemein einig, und das geht auch aus den Untersuchungen von Aterman [16] hervor. Soweit in den einzelnen Untersuchungen überhaupt eine Einschätzung der DLZ gegeben wird, steht im Zusammenhang mit dem Auftreten der DLZ und den gleichzeitig beobachteten Leberschädigungen die Auffassung im Vordergrund, daß es sich bei den DLZ um geschädigte Zellen handelt. Sie sollen Ausdruck einer beginnenden Zelldegeneration oder Formen eines senilen Zellabbaus sein, oder sie werden als erschöpfte Zellen nach vorangegangener besonderer funktioneller Belastung bzw. Überlastung angesehen [2], [5], [17], [27]—[30].

Fischler und Roeckl [3] vertreten dagegen die Ansicht, daß das Auftreten DLZ an bestimmte Funktionszustände gebunden ist. Nach ihren experimentellen Beobachtungen treten DLZ besonders in den Phasen einer intensiven Neoglykogenese auf, d. h. nach vorausgegangener Glykogenverarmung der Leber, die Fischler und Roeckl hormonell oder durch hepatotrope Gifte induzierten. Sie werten die DLZ nicht als eine degenerative Erscheinung, sondern als Ausdruck proliferativer Vorgänge in der Leber im Sinne einer Zellneubildung im Zusammenhang mit einer funktionellen Überbeanspruchung.

Auf Grund unserer eigenen Beobachtungen an der regenerierenden Rattenleber und der Leber neugeborener Ratten [20] können wir uns nicht den Meinungen anschließen, die in den DLZ Stadien einer Zelldegeneration oder eines senilen Zellabbaus sehen möchten. Wie unsere Beobachtungen ergaben, erscheinen die DLZ kurz nach der Teilhepatektomie in erhöhter Anzahl, teils weniger, teils hochgradig verdichtet. Im Verlaufe der Regeneration vermindert sich ihre Anzahl, wie aus Übersichtsaufnahmen deutlich hervorgeht. Ihr Verschwinden wäre durch zwei Möglichkeiten erklärbar: Erstens, die DLZ bilden sich allmählich in HLZ um; hierbei durchlaufen sie weniger elektronendichte Übergangsstadien, wie sie speziell 48 Std. nach Teilhepatektomie, aber auch früher und noch später neben den eigentlichen HLZ immer wieder beobachtet wurden. Zweitens ist es möglich, daß die DLZ unter den bekannten Nekroseerscheinungen degenerieren. Dagegen spricht jedoch eindeutig der Befund, daß wir innerhalb der untersuchten Zeiträume von 4 Std. bis zu 4 Tagen in keinem Fall DLZ fanden, die nekrotisch waren oder die Anzeichen einer Nekrose aufwiesen. Wir beobachteten an den DLZ auch keine Lösung der Zellverzahnungen, interzelluläre Spaltraumbildungen, Membraneinziehungen an den Mitochondrien oder ähnliche Erscheinungen, wie sie David [29] bei Mäusen nach absoluter Hungereinwirkung gesehen hat und die er als Anzeichen einer beginnenden Degeneration der DLZ bewertet. Schlicht [31] untersuchte lichtmikroskopisch die Mäuseleber nach absoluter mehrtägiger Hungereinwirkung. Hierbei beobachtete sie im Gegensatz zu einer Reihe anderer Autoren Nekrosebereiche im Zentrum der Leberazini. Sie berichtet ebenfalls über DLZ, die jedoch im Gegensatz zu anderen Befunden [2], [3], [5] nicht in vermehrter Anzahl auftraten. Inwieweit und ob überhaupt diese DLZ in Beziehung zu den Nekroseerscheinungen stehen, darüber fehlen Angaben.

Bei einem Untergang von DLZ in der regenerierenden Rattenleber hätten wir bei unseren eigenen Untersuchungen Makrophagen- oder Kupffer-Zellanschnitte mit phagozytiertem nekrotischen Material finden müssen. Das war aber niemals der Fall. All unsere Beobachtungen an der regenerierenden Rattenleber sprechen vielmehr dafür, daß die Bildung DLZ sehr wahrscheinlich mit proliferativen Vorgängen zusammenhängt, wie sie im Falle der Leberregeneration nach Teilhepatektomie oder nach Einwirkung hepatotoxischer Substanzen früher oder später immer eintreten. Unsere Auffassung wird gestützt durch den Befund, daß auch in der noch relativ rasch wachsenden Leber wenige Tage alter Ratten immer wieder einzelne typische DLZ zwischen HLZ zu finden sind (Abb. 8). Für unsere Auffassung sprechen weitere Beobachtungen [32], [33], wonach sich Zellen mit einer hohen Proliferationsrate in der Regel durch einen großen Reichtum an freien Ribosomen auszeichnen, wie das auch in den DLZ der Fall ist. Hierbei wird die Ansicht vertreten, daß derartige, mit vielen freien Ribosomen ausgestatteten Zelltypen bevorzugt Proteine für den Eigenbedarf synthetisieren (sog. retaining cells), während die an Ergastoplasma reichen Zellen den sog. secretory cells zuzuordnen sind, d. h. die neugebildeten Proteine werden größtenteils von der Zelle abgegeben. Danach würde nach Teilhepatektomie ein Teil der Leberzellen im Zusammenhang mit der Proliferation den retaining-Zelltyp durchlaufen, um später wieder den bevorzugt mit ERG-Lamellen ausgerüsteten Sekretions-Typ zu erreichen.

Völlig ungeklärt bleibt die eigentliche Ursache der hohen Elektronendichte der DLZ. Clara [2] und Böhm [28] vermuten, daß sich die DLZ infolge eines Flüssigkeitsverlustes bilden. Auch Herdson und Mitarbeiter [4] nehmen einen unterschiedlichen Hydrationsgrad des Zytoplasmas bei beiden Zelltypen an, obwohl dieser Annahme ihr Befund entgegensteht, daß der Wassergehalt von Lebern mit Hell- und Dunkelzellen nicht vom Kontrollwert abweicht. Die im Zusammenhang mit biochemischen Untersuchungen von Tsuboi und Mitarbeitern [34] durchgeführten Wassergehaltsbestimmungen an regenerierenden Mäuselebern ergaben 24 Std. nach Teilhepatektomie zwar eine Verringerung des Wassergehaltes um 3%, dieser steigt aber dann in den nachfolgenden 24 Std. sogar um fast 10% an, um erst nach etwas über 4 Wochen den Normalwert wieder zu erreichen. Da sich die DLZ im Vergleich zu den HLZ immer in der Minderzahl befinden, werden mit ihnen verbundene Änderungen im Hydratationsgrad und damit im Gesamtwassergehalt des Lebergewebes wenig zum Ausdruck kommen. Weitere definitive Aussagen zu dem Problem der DLZ und HLZ dürften überhaupt erst dann möglich sein, wenn es gelingt, durch eine saubere Zelltrennung beide Zelltypen voneinander zu isolieren, um im Falle biochemischer Analysen den Unsicherheitsfaktor des „Mischgewebes“ auszuschalten.

Wir möchten Fräulein E. Köcher für ihre gewissenhafte technische Mithilfe bei der Durchführung dieser Arbeit besonders danken

Literatur

[1] Cohn, T.: Inaug. Diss. Phys. Inst. Univ. Breslau 1892.
[2] Clara, M.: Z. mikr.-anatom. Forsch. **31**, 193 (1932).
[3] Fischler, F. u. K. W. Roeckl: Z. mikr.-anatom. Forsch. **44**, 563 (1938).
[4] Herdson, P. B., P. J. Garvin u. R. B. Jennings: Amer. J. Pathol. **45**, 157 (1964).
[5] Tauchi, H. u. T. Nakamura: Nagoya med. J. **1**, 143 (1953).
[6] Wood, R. L.: Amer. J. Pathol. **46**, 307 (1965).
[7] Gansler, H.: 5. Int. Congr. Electr. Micr., Philadelphia 1962.
[8] Hanzon, V.: Proc. Europ. Conf. Electr. Micr., Delft 1960.
[9] Hübner, G. u. W. Bernhard: Beitr. pathol. Anatom. allg. Pathol. **125**, 1 (1961).
[10] Arakawa, K.: J. Electronmicrosc. **8**, 54 (1960).
[11] Herman, L., L. Eber u. P. J. Fitzgerald: 5. Int. Congr. Electr. Micr., Philadelphia 1962.
[12] Arhelger, R. B., J. S. Broom u. R. K. Boler: Amer. J. Pathol. **46**, 409 (1965).
[13] Smuckler, E. A., O. A. Iseri u. E. P. Benditt: J. exp. Medicine **116**, 55 (1962).
[14] Thoenes, W. u. P. Bannasch: Virchow's Arch. pathol. Anatom. Physiol. klin. Med. **335**, 556 (1962).

[15] Steiner, J. W., J. S. Carruthers u. S. R. Kalifat: Z. Zellforsch. mikroskop. Anatom. 58, 141 (1962).
[16] Aterman, K.: Anatom. Rec. 136, 157 (1960).
[17] Braun, H.: Strahlentherapie 117, 134 (1962).
[18] Franke, H. u. E. Goetze: Acta biol. med. german. 14, 792 (1965).
[19] Weidemann, G., H. Franke u. E. Goetze: (In Vorbereitung).
[20] Franke, H. u. E. Goetze: (In Vorbereitung).
[21] Caulfield, J. B. : J. biophysic. biochem. Cytol. 3, 827 (1957).
[22] Kellenberger, E., W. Schwab u. A. Ryther: Experientia [Basel] 12, 421 (1956).
[23] Karnovsky, J. M.: J. biophysic. biochem. Cytol. 11, 729 (1961).
[24] Aterman, K.: J. Pathol. Bacteriol. 82, 367 (1961).
[25] Davis, J. M. G.: Acta radiol. [Stockholm] 58, 17 (1962).
[26] Trotter, B. L.: J. Cell Biol. 21, 233 (1964).
[27] Appel, J.: Med. Fak. Diss. Berlin 1953.
[28] Böhm, J.: Z. Zellforsch. mikroskop. Anatom. 15, 272 (1932).
[29] David, H.: Die Leber bei Nahrungsmangel und Mangelernährung, Akademie-Verlag, Berlin 1961.
[30] Steiner, J. W., J. S. Carruthers u. S. R. Kalifat: Exp. mol. Pathol. 1, 162 (1962).
[31] Schlicht, I.: Virchow's Arch. pathol. Anatom. Physiol. klin. Med. 330, 436 (1957).
[32] Birbeck, M. S. C. u. E. H. Mercer: Nature [London] 189, 558 (1961).
[33] Howatson, A. F. u. A. W. Ham: Canad. J. Biochem. Physiol. 35, 549 (1957).
[34] Tsuboi, K. K., H. O. Yokoyama, R. E. Stowell u. M. E. Wilson: Arch. Biochem. Biophysics 48, 275 (1954).

Summary

H. Franke and E. Goetze: Electron-microscopic studies on „light" and „dark" liver cells in the regenerating rat liver

The substructure of the rat liver was investigated at 4 to 96 hours following partial hepatectomy. Regeneration is accompanied by characteristic fine-structural changes. Initially, the liver becomes intensively fatty, symptoms of vacuolation occuring in the endoplasmic reticulum, accompanied by a general reduction of the ergastoplasm and a simultaneous occurrence of numerous free ribosomes.

Especially striking is the presence of an increased number of dark liver cells amongst the light ones, which are characterized by a very high electron density and a high content of free ribosomes. Whereas the fine-structural changes in the light liver cells are restored rather rapidly, the characteristic substructure in the dark liver cells remains unchanged. They are regressing very slowly without any symptoms of degeneration. Electron-microscopic data from the regenerating liver and those of newborn rat suggest that the dark liver cells are closely related with an increased cell proliferation, and that they may turn again to light cells during regeneration.

Резюме

Г. Франке и Е. Гётце: Электронномикроскопические исследования о „ярких и темных" печеночных клетках в регенеривующей печени крысы

Исследовалась субструктура печени крысы через 4 до 96 часов после частичной гепатектомии. Процесс регенерации сопровождался характерными изменениями тонкой структуры. Вначале наблюдали равномерное интенсивное ожирение печени и явления вакуолизации на эндоплазматическом ретикулуме, сопровождающиеся общим уменьшением эргастоплазмы с одновременным появлением многочисленных свободных рибосом.

Особенный интерес вызывает наличие большого числа темных печечных клеток наряду с яркими, отличающихся высокой электронной плотностью, а также высоким содержанием свободных рибосом.

В то время как в ярких печечных клетках изменения тонкой структуры нормализуются довольно быстро, в темных клетках сохраняется их характерная субструктура. Количество их уменьшается лишь медленно, причем признаков дегенерации установить не удалось.

Электронномикроскопические данные регенерирующей печени и печени новорожденных крыс указывают на то, что появление темных печеночных клеток связано с повышенной пролиферацией клеток и что темные клетки могут превращаться в яркие в процессе регенерации.

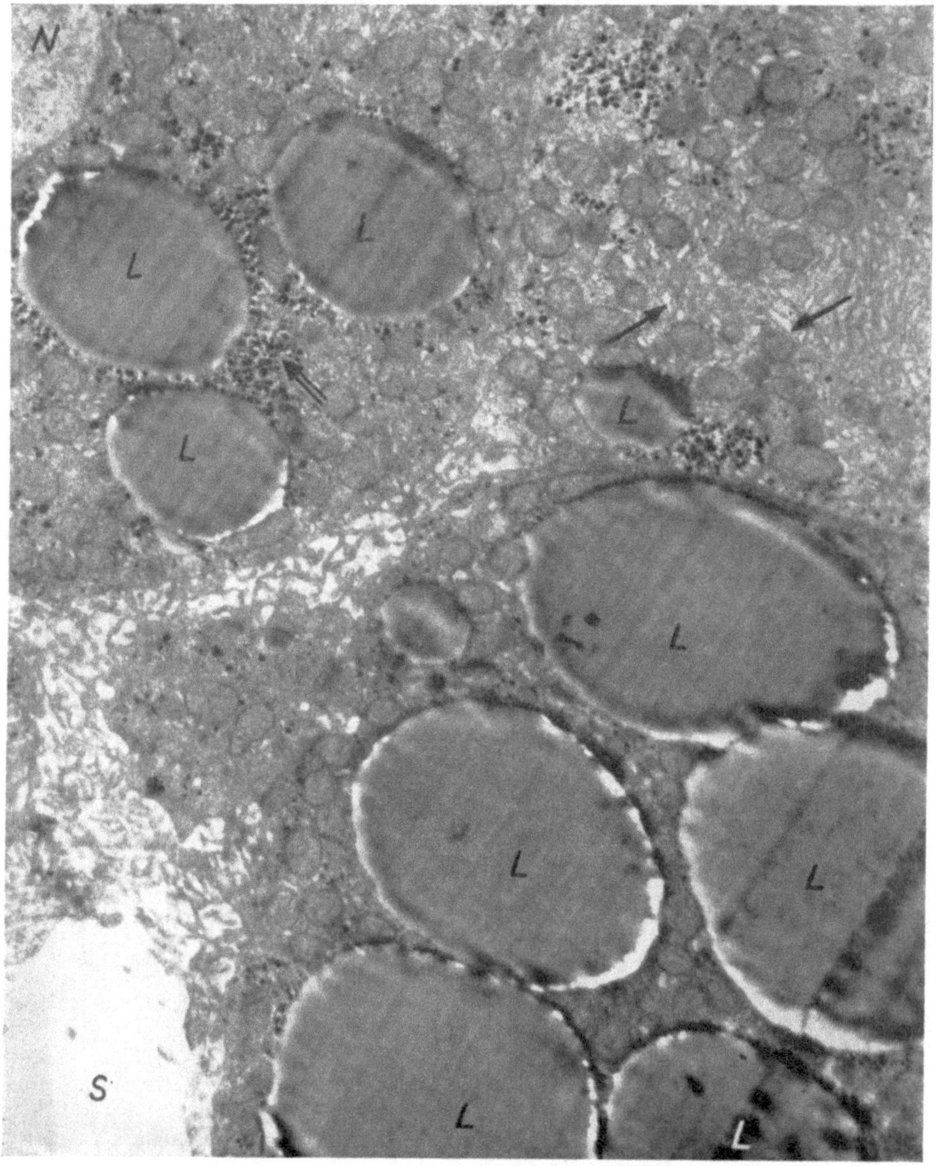

Abb. 1. Rattenleber 4 Std. nach Teilhepatektomie

Lipideinschlüsse (L); das Ergastoplasma zeigt teilweise Vakuolisierungserscheinungen (↗); Zellkern (N); glykogenhaltige Partikel (⇒); Sinusoid (S).
Gesamtvergr.: 9600fach

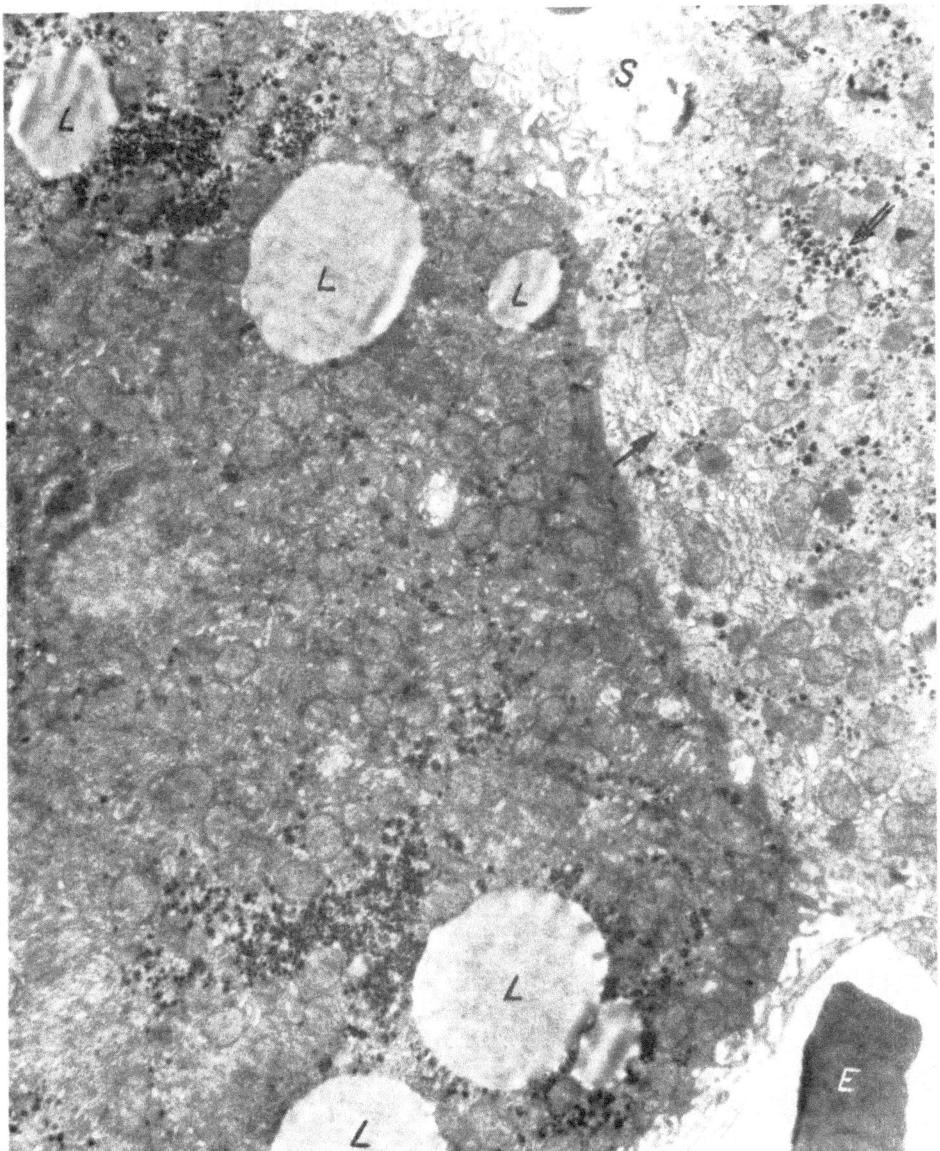

Abb. 2. Typische dunkle Leberzelle neben einer hellen 4 Std. nach Teilhepatektomie. Vakuolisiertes ER (↗); glykogenhaltige Partikel (⇒); Lipideinschlüsse (L); Erythrozyt (E); Sinusoid (S).

Gesamtvergr.: 9600fach

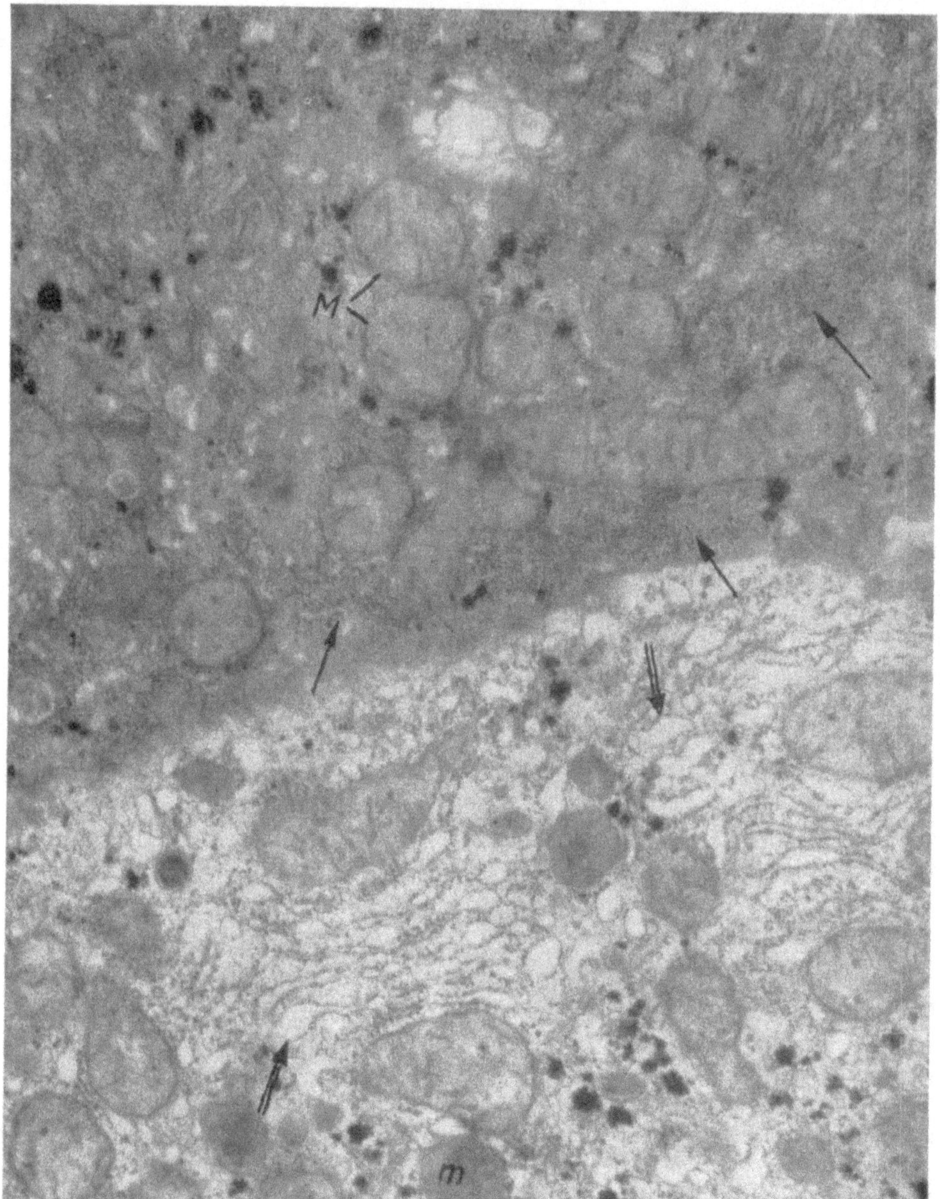

Abb. 3. Bereich einer hellen und dunklen Leberzelle 4 Std. nach Teilhepatektomie
ER in der hellen Zelle vakuolisiert (⇒); in der dunklen Zelle zahlreiche freie Ribosomen (↗); Mitochondrien (M); Mikrobody (m).
Gesamtvergr.: 23 500fach

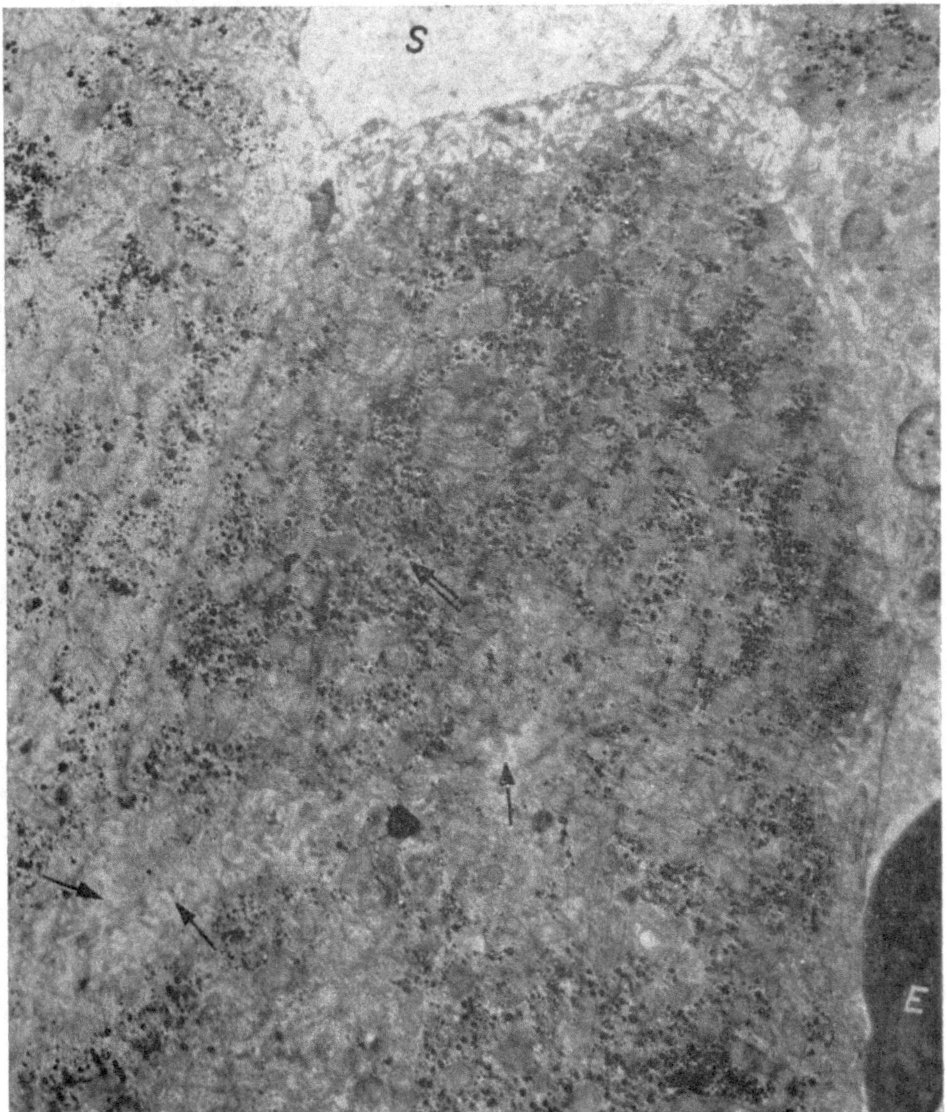

Abb. 4. Rattenleber mit hellen und dunklen Leberzellen 24 Std. nach Teilhepatektomie. Die vorher so zahlreichen Lipideinschlüsse sind verschwunden, ebenso die Vakuolisierungserscheinungen, Ergastoplasmamembranen sind nur spärlich entwickelt.

Glykogenhaltige Partikel (⇑); Erythrozyt (E); Gallenkapillaren (↗); Sinusoid (S).
Gesamtvergr.: 8000fach

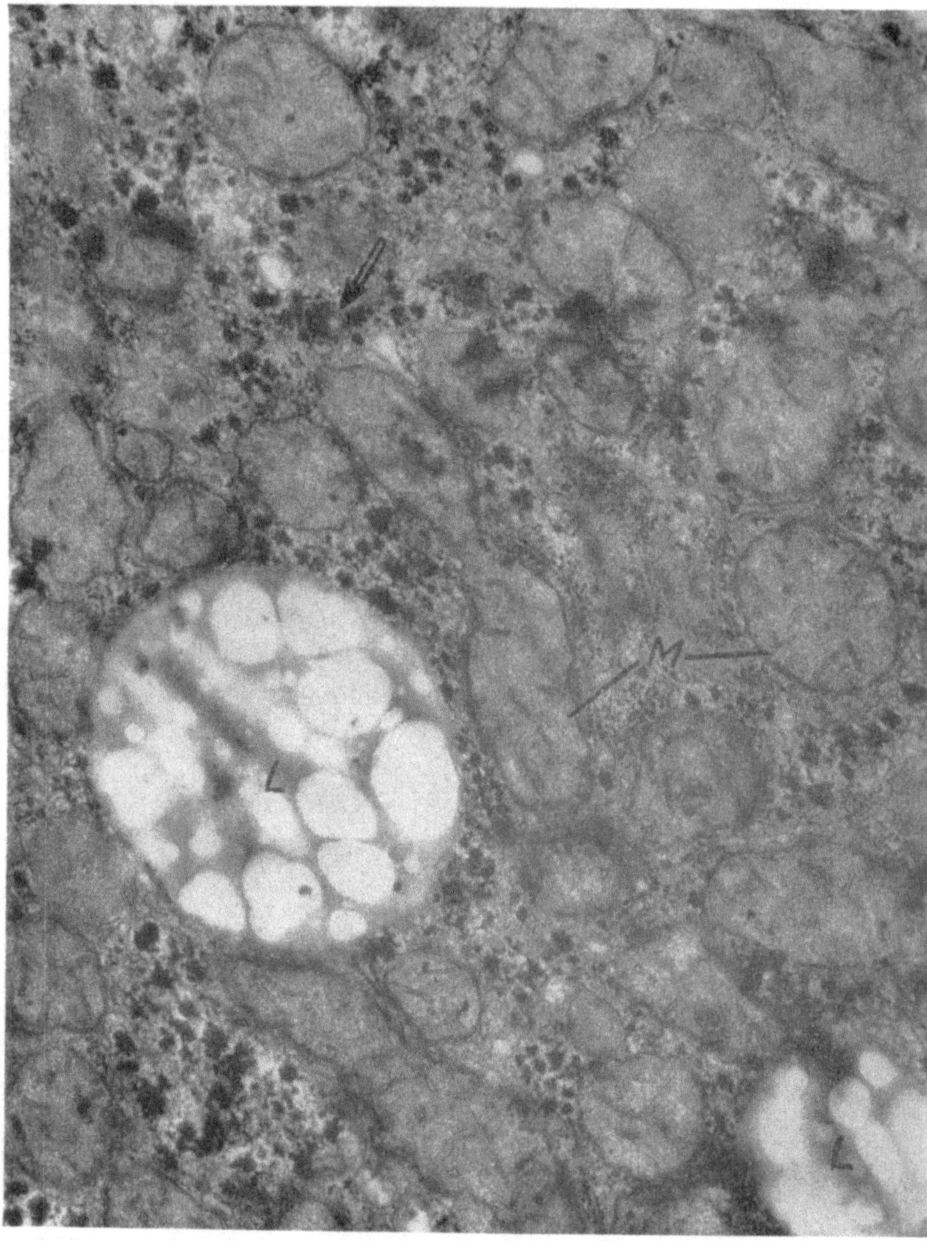

Abb. 5. Dunkle Leberzelle 24 Std. nach Teilhepatektomie.
Lipideinschlüsse (L); Mitochondrien (M); glykogenhaltige Partikel (⇒).
Gesamtvergr.: 39200fach

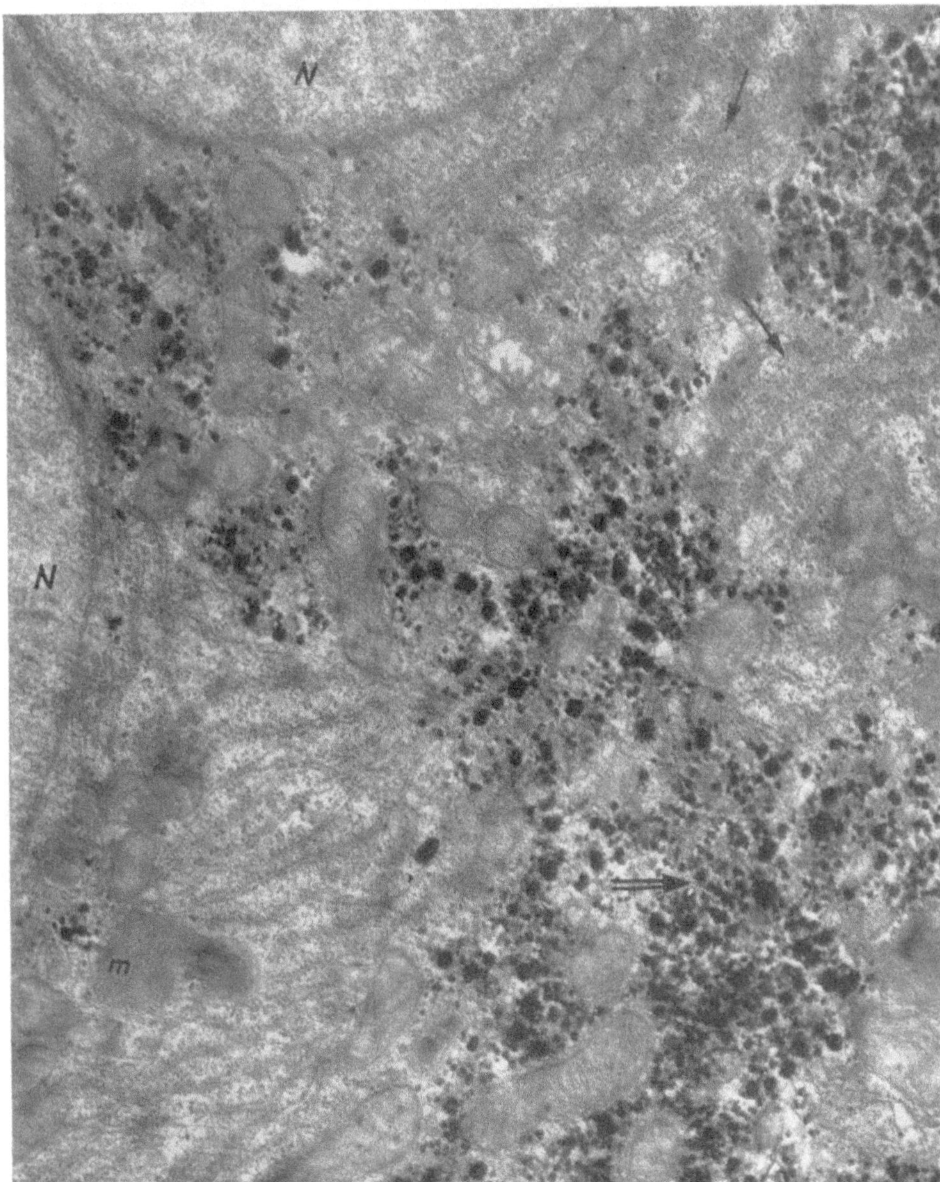

Abb. 6. Zweikernige Leberzelle 48 Std. nach Teilhepatektomie.

Ergastoplasmamembranen gut mit Ribosomen besetzt, daneben auch zahlreiche freie Ribosomen (↗); Zellkern (N); glykogenhaltige Partikel (⇒); Mikrobody (m).

Gesamtvergr.: 17 500fach

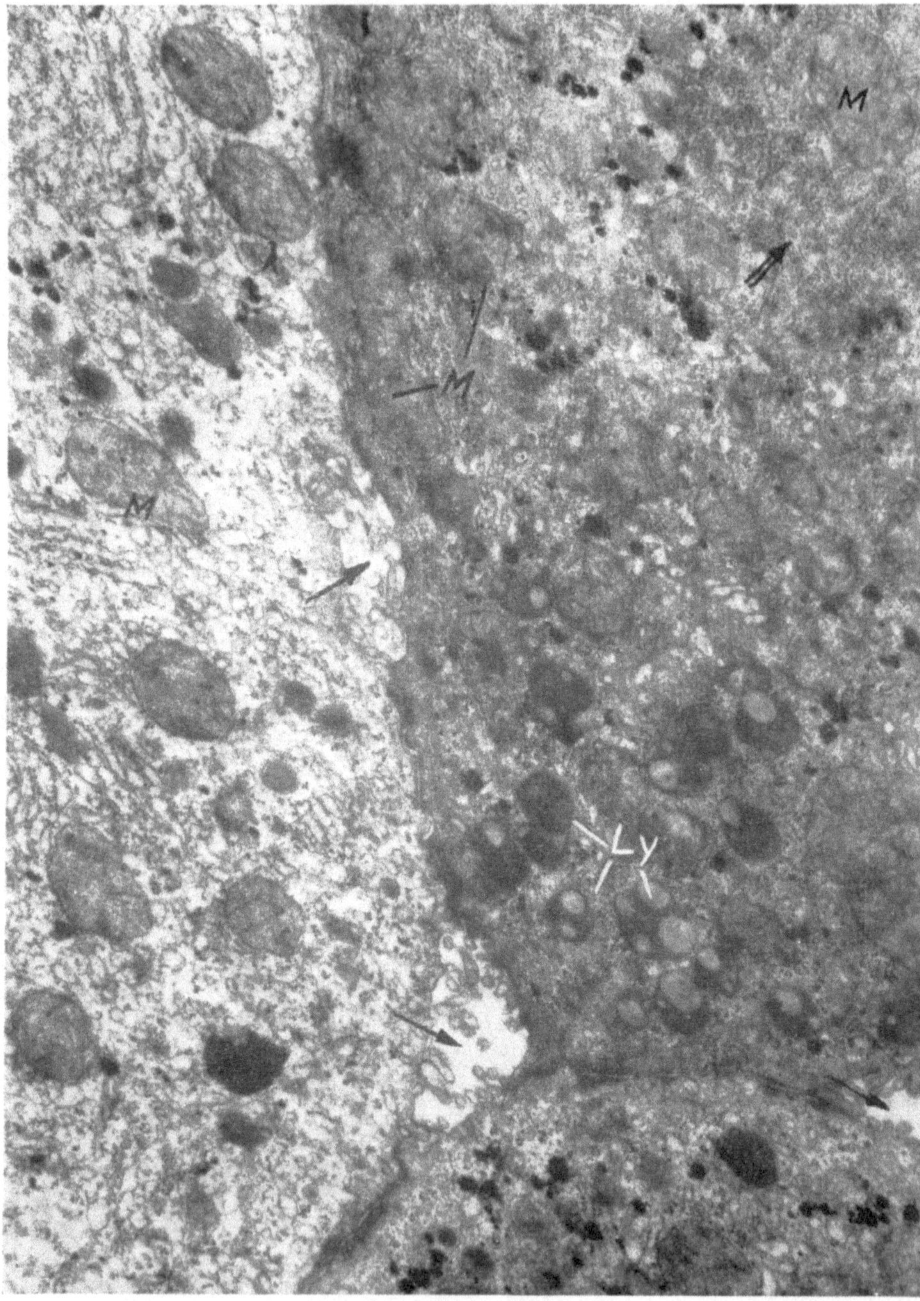

Abb. 7. Bereich einer hellen und dunklen Leberzelle 96 Std. nach Teilhepatektomie. Mitochondrien (M); Lysosomen (Ly); Ansammlung freier Ribosomen (⇒); Gallenkapillare (↗).
Gesamtvergr.: 17 500fach

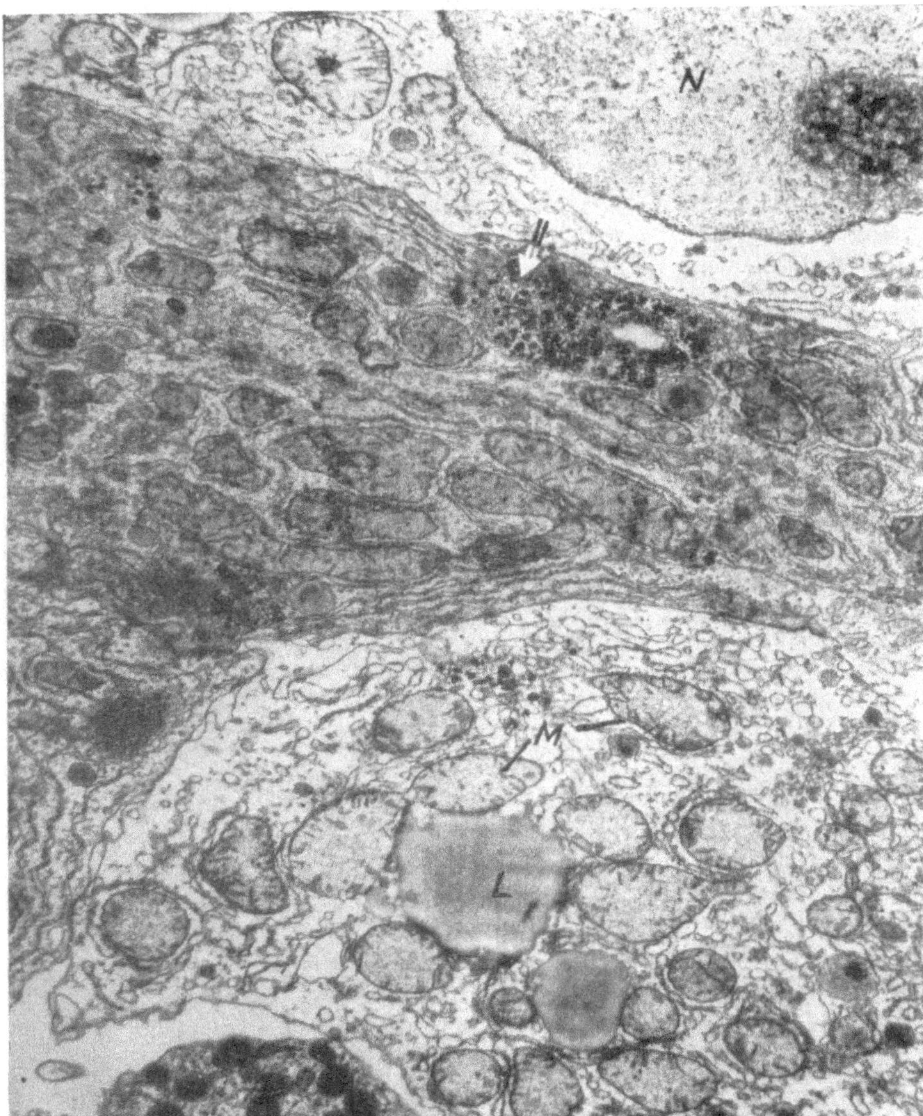

Abb. 8. Normale Leber einer 28 Std. alten Ratte mit hellen und dunklen Zellen. Zellkern (N); Lipideinschluß (L); Mitochondrien (M); glykogenhaltige Partikel (⇒). Gesamtvergr.: 21000fach

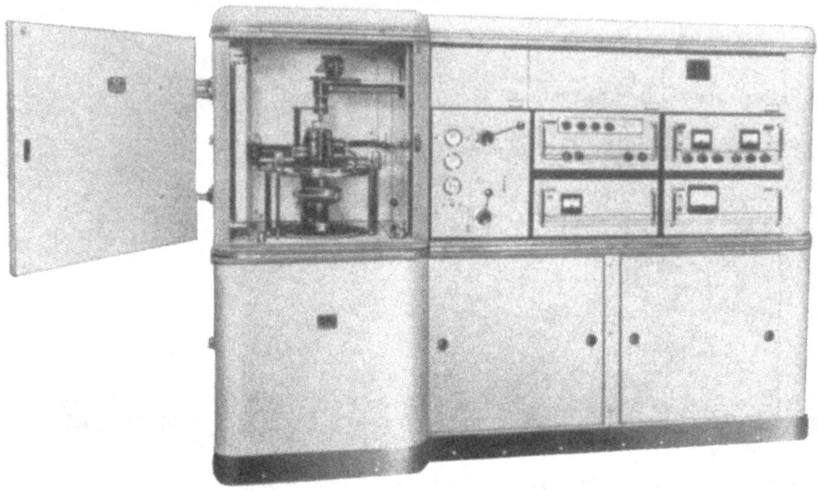

CONTENTS

СОДЕРЖАНИЕ

INHALT

Biochemie

Herausgeber und verantwortlich für den Inhalt. R. Baumann, H. Dutz, A. Graffi, H. Gummel, F. Jung, L.-H. Kettler, S. M. Rapoport. Schriftleitung Prof. Dr. Werner Scheler und Dr. habil. Heinz Bielka, 1115 Berlin-Buch, Lindenberger Weg 70. Fernruf: 56 98 51. App. 222. Verlag: Akademie-Verlag GmbH, 108 Berlin, Leipziger Straße 3—4 (Fernruf: 22 04 41) Telex-Nr. 011 773, Postscheckkonto Berlin 3 50 21. Bestellnummer dieses Heftes 1053/XVII/1. Die Zeitschrift erscheint monatlich. Bezugspreis: Einzelheft MDN 14,—, Doppelheft MDN 28,—. Veröffentlicht unter der Lizenznummer 1318 des Presseamtes beim Vorsitzenden des Ministerrates der Deutschen Demokratischen Republik. Gesamtherstellung: VEB Druckerei „Thomas Müntzer", 582 Bad Langensalza.

FSC
www.fsc.org
MIX
Papier | Fördert gute Waldnutzung
FSC® C083411

Zeitfracht Medien GmbH
Ferdinand-Jühlke-Straße 7
99095 Erfurt, Deutschland
produktsicherheit@kolibri360.de